Advanced Metaheuristic Approaches for Graph and Scheduling Optimization

George
Mills

Contents

Chapter 1

Introduction

Combinatorial Optimization Problems (COPs), rooted in combinatorics, operations research, and theoretical computer science, are ubiquitous across science, engineering, and industry. Fundamentally, a COP seeks to identify an optimal solution from a large, yet finite, set of discrete feasible solutions, all while adhering to specified constraints.

Despite their often simple statement, the intrinsic challenge of COPs lies in their inherent computational complexity, driven by what is termed "combinatorial explosion" – the exponential growth in the number of possible solutions as the size of a problem instance increases. For example, in the traveling salesperson problem (TSP), even a small increase in the number of cities leads to a dramatic rise in the number of potential routes to evaluate. Due to this inherent property, finding an optimal solution to most of such problems, in practice, are really hard.

While combinatorial optimization is often considered a relatively young discipline in its formalized structure, its conceptual roots extend deep into antiquity. The foundational understanding of combinatorial principles applied to finite discrete objects has been present in human thought for millennia. Many classic COPs emerged directly from practical necessities, such as determining the shortest route between two points or planning efficient tours. In earlier times, such problems were approached individually, often through ad hoc mathematical methods.

A pivotal moment came in 1736 with Leonhard Euler's solution to the *Seven Bridges of Königsberg* problem. This groundbreaking contribution not only laid the groundwork for graph theory, a new branch of mathematics, but also critically facilitated the eventual development and formalization of combinatorial optimization as a distinct field. Further formalization occurred through the development of linear programming, particularly with the contributions of Leonid Kantorovich in 1939 and George Dantzig in 1947, whose simplex method became a cornerstone of optimization by the 1950s.

The advent of digital computers revolutionized problem-solving but also revealed a fundamental limitation: many COPs remained intractable. While some problems, like the shortest path, can be solved efficiently with polynomial-time algorithm, others resist such solutions. This is because the number of potential solutions explodes exponentially as the problem size grows, making an exhaustive search computationally infeasible. Despite extensive research efforts, finding a polynomial-time algorithm for these more intricate COPs remains elusive. This challenge gave rise to the theory of computational complexity. Seminal work by Stephen Cook and Richard Karp in the 1970s introduced and formalized the notions of $\mathcal{NP}$-completeness and $\mathcal{NP}$-hardness, providing a theoretical basis to distinguish between problems that can be solved efficiently and those that are believed to be intractable.

The rapid pace of innovation in fields like telecommunications and circuit design, coupled with the emergence of new areas such as social and biological networks, has led to a proliferation of these challenging $\mathcal{NP}$-hard COPs. Despite significant advances in computing power, efficiently solving such problems continues to be a formidable and central challenge for researchers worldwide.

1.1 Combinatorial optimization problems

A combinatorial optimization problem (COP) is combinatorial in nature and involves finding an optimal solution from a finite, yet potentially very large, set of feasible solutions. Each solution must satisfy specific constraints, and the objective is to either minimize or maximize a given function.

Formally, a combinatorial optimization problem Π for a given instance can be represented as a triple (X, F, Ω), where:

- X is the set of all feasible solutions that satisfy the given constraints. This set is also referred to as the *search space*.

- F is the objective function that assigns a numerical value $F(s)$ to each solution $s \in X$.
- Ω is the set of constraints that determine the feasibility of a solution.

A solution that satisfies all the constraints in Ω is called a *feasible solution.* A key characteristic of a COP is that the variables involved in constructing a solution are discrete. The goal of the problem Π is to find a feasible solution s' that optimizes the objective function.

- In a minimization problem, the objective is to find a solution $s' \in X$ such that

$$F(s') \leq F(s), \quad \forall s \in X.$$

- In a maximization problem, the objective is to find a solution $s' \in X$ such that

$$F(s') \geq F(s), \quad \forall s \in X.$$

It is possible that multiple solutions may have the same optimal objective value. In this thesis, the notations F and f will be used interchangeably to represent the objective function. The function ($F(s)$ or $f(s)$) evaluates the objective value – also referred to as the quality or fitness – of a solution s, thereby guiding the search process.

COPs are not merely abstract mathematical constructs; they play a vital role in real-world decision-making across diverse domains, including logistics and transportation, computer science and engineering, operations research, telecommunications, and biology.

Due to their structural characteristics, COPs can typically be classified into three distinct categories. This classification is helpful for analyzing the problem structure and developing appropriate solution strategies:

1. Subset problems: These problems involve selecting a subset of items from a larger set to optimize an objective function, subject to given constraints. For examples, set covering problem, vertex cover problem, and minimum spanning tree problem.
2. Permutation problems: these problems involve finding an optimal sequence or arrangement of elements to optimize an objective function, subject to given constraints. For example, traveling salesman problem, and job scheduling problem.

3. Grouping problems: these problems involve assigning all items to groups in a way that optimizes an objective function, subject to given constraints. For example, bin packing problem, and graph coloring problem.

It is worth emphasizing that a single combinatorial optimization problem (COP) may embody features from multiple structural categories. Moreover, the majority of COPs are classified as $\mathcal{NP}$-hard, indicating that the computational time required by any known exact method increases exponentially with the size of the problem instance. This exponential growth, driven by the combinatorial explosion of feasible solutions, makes exact methods impractical for all but the smallest problem instances – even in the face of rapid advancements in computing power.

Despite decades of intensive research, no deterministic polynomial-time algorithm has been discovered for any $\mathcal{NP}$-hard problem. Due to their inherent complexity, these problems render exact methods impractical for solving instances of significant size.

As a result, researchers are compelled to turn to approximate methods [1], which prioritize solution quality over guaranteed optimality. Examples of such methods include problem-specific heuristics, approximation algorithms, and metaheuristic techniques. Among these, metaheuristic techniques have gained significant prominence due to their ability to find high-quality, near-optimal solutions for $\mathcal{NP}$-hard COPs within a reasonable amount of computational time. These approaches are particularly valuable for large and complex problem instances where exact methods become impractical.

It is important to note that metaheuristic techniques often return solutions with objective values that are close to optimal; however, such solutions may be far from the true optimum in the search space. This is because the search space of an $\mathcal{NP}$-hard problem is typically rugged, characterized by a vast number of local optima – some of which have objective values nearly equal to that of the global optimum.

1.2 Metaheuristic techniques

A metaheuristic technique is a high-level problem-solving method designed to guide underlying heuristics in navigating complex and large search spaces. These techniques are particularly effective for solving complex optimization problems where traditional exact methods are computationally infeasible. Although they don't guarantee finding an optimum solution, metaheuristic techniques are capable of finding high-quality, near-optimal solutions within a reasonable amount of computational time in the search space [1, 2]. A key strength of these methods lies in their flexibility; they can be adapted to a wide range of optimization problems with minimal modifications. Their adaptability makes them especially suitable for $\mathcal{NP}$-hard combinatorial optimization problems across various domains.

Several metaheuristic approaches have gained significant attention due to their effectiveness for solving complex combinatorial optimization problems, particularly those classified as $\mathcal{NP}$-hard. These include population-based methods such as genetic algorithms, ant colony optimization, and artificial bee colony (ABC) algorithm, as well as single solution-based methods like tabu search, iterated local search, and variable neighborhood search. Each technique employs distinct strategies for balancing exploration and exploitation, enabling them to effectively navigate the search space and avoid becoming trapped in local optima.

The foundational elements in designing a robust and effective metaheuristic technique tailored to a specific $\mathcal{NP}$-hard combinatorial optimization problem include the solution encoding (solution representation), the objective function to evaluate the quality of solutions, and constraint-handling mechanisms to ensure feasibility throughout the search. Importantly, the design of search operators – such as genetic/neighborhood /perturbation/shaking or local search – depending on the underlying principles of the metaheuristic. These operators play a crucial role in generating new candidate solutions and guiding the search process through a balance of exploration and exploitation.

In this context, solution encoding for a particular problem refers to how to encode (represent) a solution so that the metaheuristic technique should converge faster towards high quality solutions. A metaheuristic technique alternates between the encoding space and the solution space. When defining a solution encoding, the solution should be evaluated through a suitable evaluation function in such a way that the notion of distance in solution space should be preserved in encoding space, and it should be compatiable with the operators (such as neighborhood operators, local search operators). In addition, A well-designed metaheuristic technique must maintain a balance between exploring new regions of the search space and exploiting local area around the current solution. Too much exploration may lead to slow convergence, while too much exploitation can result in getting trapped in poor local optima.

Researchers are often involved in the design and refinement of metaheuristic techniques tailored to the characteristics of specific problems. These methods are rigorously tested on benchmark data-sets and compared with established state-of-the-art methods to evaluate their performance in terms of both solution quality and computational time. Depending on the problem under consideration and the methodological advancements introduced, the resulting method may emerge as an improved variant of an existing metaheuristic approach, an entirely novel approach, or a hybrid approach that combines multiple strategies to leverage their combined strengths.

The effective design of metaheuristic techniques requires a deep understanding of both the problem structure and the search dynamics of the selected approach. Incorporating problem-specific knowledge into the components of a metaheuristic technique often helps achieve higher solution quality with lower computational effort.

This thesis focuses on three metaheuristic techniques – artificial bee colony (ABC) algorithm, iterated local search (ILS), and variable neighborhood search (VNS) – tailored to four different $\mathcal{NP}$-hard COPs. Among them, ABC algorithm is a population-based metaheuristic technque belonging to the class of swarm intelligence techniques, while ILS and VNS are categorized as single-solution-based metaheuristic techniques.

1.2.1 Swarm intelligence technique

Swarm Intelligence (SI) is a research paradigm inspired by the collective behavior of decentralized, self-organized systems composed of simple agents such as bees, ants, birds, or fish. These agents interact locally with one another and their environment through simple rules, leading to the emergence of intelligent global behavior without centralized control. The term "swarm intelligence" was introduced in the context of cellular robotic systems [3]. SI is characterized by two key features: *self-organization* and *division of labor*.

Self-organization arises from local interactions among agents and comprises four key elements [4]:

- Positive feedback: Reinforces beneficial behaviors, such as recruitment dances performed by bees.
- Negative feedback: Acts as a stabilizing mechanism through processes like food

source exhaustion in the course of foraging by bees.

- Fluctuations: Introduce randomness that enables the swarm to explore new opportunities or solutions.
- Multiple interactions: Allow for the dissemination of information throughout the system, supporting collective decision-making.

Division of labor means that different agents in the swarm can perform different tasks at the same time. This characteristic is especially evident in honey bee colonies.

1.2.1.1 Behavior of natural honey bees

The behavior of natural honey bees has been extensively studied and has inspired various computational models. Bees are social insects that live and work collectively within a colony. They exhibit a high level of intelligence, often participating in collective decision-making processes to solve complex tasks. Equipped with advanced navigation systems and strong memory capabilities, bees divide labor among colony members to perform specialized roles such as foraging, communication, nest maintenance, and defense.

In a honey bee colony, the queen bee is responsible for reproduction, laying both fertilized and unfertilized eggs. Fertilized eggs develop into female worker bees, while unfertilized eggs become male drones. The queen mates only once in her lifetime but can store sperm to fertilize eggs for several years. After mating, drones die.

Worker bees carry out a wide range of duties depending on their age, including constructing wax combs, cleaning and ventilating the hive, guarding the entrance, and foraging for nectar. Among these, foraging is considered one of the most vital tasks. Forager bees travel long distances to locate food sources and then return to the hive to share information about the location and quality of the nectar.

The sophisticated and decentralized behaviors exhibited by honey bees, particularly their foraging and task allocation strategies, have drawn significant attention from researchers. Numerous computational approaches have been developed to mimic various aspects of bee behavior, from queen dynamics [5, 6, 7] and mating/reproduction [8, 9, 10, 11] to, most notably, the foraging patterns of honey bees [12, 13, 14, 15, 16]. These studies have led to a rich array of algorithmic concepts. Comprehensive surveys on numerous computational approaches inspired by bee behavior can be found in [17, 18]. Among these, artificial bee colony (ABC) algorithm, introduced by Karaboga [16], has garnered significant attention for addressing complex optimization problems across a variety of domains [19, 20, 21] since its inception.

This study specifically focuses on artificial bee colony (ABC) algorithm, drawing inspiration from the intelligent foraging behavior of natural honey bees. Figure 1.1 explains this natural foraging behavior in detail.

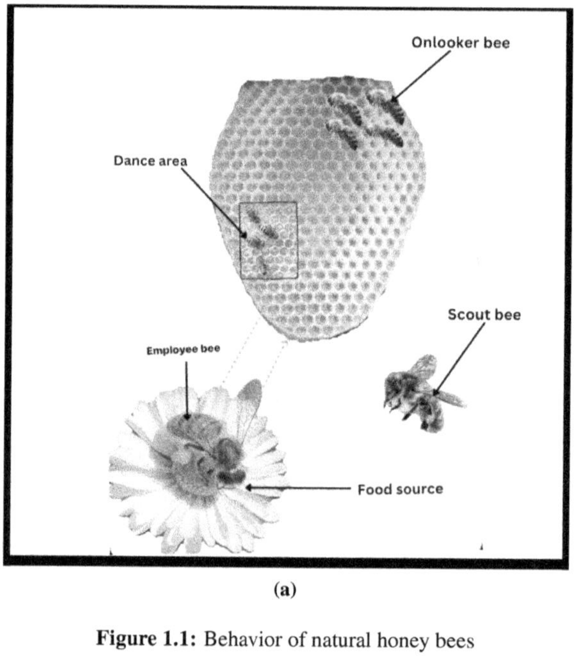

(a)

Figure 1.1: Behavior of natural honey bees

Figure 1.1 explains how natural honey bees search for and collect food, a behavior known as foraging. During foraging, some bees (called scout bees) leave the hive and fly long distances to discover new food sources, such as flowers rich in nectar. Once a food source is discovered, these bees become employed bees and collect information about the food source like the amount of nectar available, the distance from the hive to the food source, and its direction from the hive. The employed bees then collect nectar from the food source and store it in their stomachs. While in their stomachs, the nectar starts to mix with enzymes, starting the process of turning it into honey. When they return to the hive, the employed bees unload the nectar into the hive's empty cells. After unloading the nectar, the employed bee communicates with other worker bees (called onlooker bees) to recruit them for collecting nectar. This communication occurs through most interesting form of communication – a dance called "waggle dance", which tells onlooker bees about the location of the food, the quality of the food. The direction, distance, and quality of the food are shared through symbolic movements and vibrations during the dance.

1.2.1.2 Artificial bee colony algorithm

Artificial bee colony (ABC) algorithm is a swarm-based metaheuristic technique, drawing inspiration from the foraging behavior of natural honey bees. Karaboga [16] proposed an artificial bee colony (ABC) algorithm for tacking complex optimization problems. In ABC algorithm, artificial bees, as per their task, are classified into three groups – employed bees, onlooker bees, and scout bees – work in cooperation to find high-quality solutions within the search space. In ABC algorithm, a food source refers to a feasible solution to the optimization problem, and the nectar amount of a food source corresponds to the fitness or quality of that solution. ABC algorithm presumes that each employed bee is associated with a specific food source which means the number of employed bees is equal to the number of food sources. The size (say E_{pop}) of this set is referred to as the employed bee population. Typically, the number of onlooker bees is also set to equal to the number of employed bees. The size (say On_{pop}) of this set is referred to as the onlooker bee population. When a food source is no longer viable, the associated employed bee becomes a scout bee. The scout bee then explores the search space randomly to discover a new food source. Once a new food source is discovered, this scout bee reverts to the role of an employed bee and starts exploiting the new food source. In this thesis, the terms food source and solution are used interchangeably within ABC algorithm framework.

Being a population-based metaheuristic technique, ABC algorithm starts with a set of generated initial solutions, referred to as food sources. Subsequently, ABC algorithm iteratively (generation) progresses through three phases – employed bee phase, scout bee phase, and onlooker bee phase. Each group of artificial bees plays a specific role in the search process:

- Employed Bee Phase: During this phase, each employed bee generates a new solution in the neighborhood of its currently associated solution. If this new solution is superior to its current solution, the associated employed bee abandons its current solution and moves to this new solution; otherwise, the associated employed bee continues with the current solution.

- Scout Bee Phase: In this phase, if a solution ceases to show improvement for a specific number of generations – determined by the limit parameter – the employed bee associated with it promptly abandons its non-improving solution, transitioning its status to that of a scout bee. The scout bee then randomly explores the search space to discover a new solution. Once a new solution is discovered, the scout bee reverts to the state of an employed bee.

- Onlooker Bee Phase: Once the employed bee phase is completed, the job of each onlooker bee starts. Each onlooker bee selects a solution associated by an employed bee, typically via a probability-based selection method which mimics the selection of food source whose information is shared by employed bee through waggle dance. This mechanism increases the likelihood of selecting higher-quality solutions. Each onlooker bee, similar to the job of generating a new solution by an employed bee in the employed bee phase, generates a new solution in the neighborhood of its selected solution. Once, all onlooker bees complete the job of generating new solutions, the newly generated solutions –created in the neighborhood of a particular employed bee solution (say X) selected by one or more onlooker bees – along with the original employed bee solution X, compete for the position of X in the next iteration. The best among them will be selected to occupy the new position of X in the next iteration. After the new positions for all solutions are determined, the next iteration of ABC algorithm carries out.

In ABC algorithm, solution encoding is used to represent a solution of a particular problem as a food source.

The actual process for generating a new solution in the neighborhood of a particular solution (i.e., neighborhood operator) relies on the nature of the problem under consideration. In this thesis, the process of determining a new neighboring solution (neighborhood operator) for a particular problem will be discussed in detail in chapters 2 to 5.

The roulette wheel selection method and the binary tournament selection method that are the two most commonly used selection methods are used for the selection of a new solution for an onlooker bee.

In the roulette wheel selection method, each individual (solution) is assigned a selection probability that is proportional to its relative fitness. If $f(i)$ is the fitness of solution i in the population (here population is the number of employed bee solutions) whose population size is e_{pop}, then its probability (p_i) of being selected is

$$p_i = \frac{f(i)}{\sum_{j=1}^{e_{pop}} f(j)} \tag{1.1}$$

In the binary tournament selection method, two individuals are uniformly selected at random from the current population. With a given probability (say P_{bt}), better one between two selected individuals is selected, otherwise the worse one is selected with the probability (1-P_{bt}). This method is widely used, as it is computationally efficient.

Algorithm 1 illustrates a general framework of ABC algorithm.

Algorithm 1: A general framework of ABC algorithm

Generate E_{pop} solutions randomly viz. $E_1, \ldots, E_{E_{pop}}$; `// ` E_{pop} `is the number of employed bees`

best-so-far ← Best solution in E_{pop} solutions;

while *Termination criteria is not met* **do**

    Employed Bees Phase;

    Scout Bees Phase;

    Onlooker Bees Phase;

return *best-so-far*;

ABC algorithm was initially devised for continuous optimization problems [16]. In the domain of combinatorial optimization problems, Singh [20] was the first to propose a variant of ABC algorithm that is applicable for the leaf-constrained minimum spanning tree problem. Later, literature has witnessed successful applications of ABC algorithm to various combinatorial optimization problems [19, 21].

1.2.2 Iterated local search

Iterated Local Search (ILS) is a single solution based metaheuristic technique for tackling $\mathcal{NP}$-hard combinatorial optimization problems through navigating a walk through the search space. ILS has witnessed high performance with numerous $\mathcal{NP}$-hard combinatorial optimization problems [22, 23]. The pseudo-code of a general framework of ILS is illustrated in Algorithm 2. It starts with a randomly generated initial solution (say S), and then iteratively refines it over multiple iterations to discover solutions of progressively better quality in the search space. In each iteration, ILS applies a sequence of three main procedures (perturbation, local search and acceptance criterion) that help in finding solutions of high quality in the search space. The general procedures of ILS are as follows:

- **Perturbation:** The perturbation procedure modifies the current solution (S) to escape from its local optimum, producing a new solution (S'). This step introduces controlled randomness or structural change, allowing the search to move into a new region of the serach space.

- **Local search:** A local search refers to a procedure that starts from a given solution and explores its neighborhood by making small, localized changes to iteratively improve the solution. The process continues until a local optimum is reached. Here a local search is applied to the perturbed solution S', resulting in a new locally optimal solution S''. This phase intensifies the search around S' by exploiting its neighborhood to identify high-quality solutions.

- **Acceptance Criterion:** In its basic form, ILS follows a simple "better-only" acceptance rule. If the newly obtained local optimum solution (S'') is better than the current solution (S), S'' is selected as the starting point for the next iteration, and the walk in the search space continues from there. Conversely, if S is better, it remains the starting solution for the next iteration. However, ILS is flexible to adopt various acceptance criteria for enhancing exploration. For example, ILS can be restarted from a new solution if no improvement is observed over a predefined number of iterations. Roughly, the acceptance criterion helps control the balance between intensification and diversification in the search process.

The simple idea of ILS has been independently rediscovered by many authors, leading to various names for the approach, such as iterated descent [24, 25], large-step Markov

Algorithm 2: The pseudo-code of general framework of ILS

```
Generate an initial solution S;
while The termination criterion is not met do
    S' ← Perturbation(S);
    S' ← LocalSearch(S');
    S' ← AcceptanceCriterion(S, S*');
```

chains [26], iterated Lin-Kernighan [27], chained local optimization [28], or combinations of these [29]." Lorenço *et. al* [23] highlighted two points that define ILS: (i) It explores the search space by following a single chain of solutions, meaning it does not rely on a population of solutions as in population-based methods. (ii) The search for better solution takes place in a reduced search space, shaped by the output of an embedded heuristic (commonly a local search procedure). Although local search is the most commonly used embedded heuristic in ILS, the framework is flexible and can incorporate any optimization method – deterministic or stochastic – as long as it produces locally optimal solutions. This flexibility, combined with its simplicity and ease of implementation, makes ILS a general-purpose framework that can be adapted to a wide range of combinatorial optimization problems.

1.2.3 Variable neighborhood search

Variable Neighborhood Search (VNS) is a single solution-based metaheuristic technique [30] that has shown promising results for various $\mathcal{NP}$-hard combinatorial optimization problems [31, 32, 33]. Mladenović and Hansen [30] introduced [30] VNS as a metaheuristic technique that systematically changes the neighborhood structures during the search process to escape local optima and improve solution quality. VNS is based on the following facts that explain the use of multiple neighborhood structures which can be highly effective in avoiding local optima and improving overall search quality:

1. A local optimum in one neighborhood is not necessarily a local optimum in another. This fact is exploited by gradually applying more complex neighborhood changes, aiming to find local optima with respect to all neighborhood structures used.

2. A globally optimal solution is a local optimum with respect to all possible neighborhood structures.

3. In many problems, local optima found using different neighborhood structures tend to lie near each other in the search space.

VNS does not follow a trajectory in the search space. Instead it explores increasingly distant neighborhoods of the current incumbent solution, and moves to a new one only if a better solution is identified. Additionally, some local search method is applied at each step to refine these neighboring solutions and reach local optima.

Algorithm 3: The pseudo-code of a Basic VNS framework

```
Initialize with generating a random solution S;
Best ← S;
while termination criterion is not fulfilled do
    for k ← 1 to K_max do
        S′ ← Shaking(S, N_k);
        S″ ← LocalSearch(S′);
        if the fitness of S″ is better than S then
            S ← S″;
            k ← 1;
            if the fitness of S is better than Best then
                Best ← S;
        else
            k ← k + 1;
```

The Basic VNS framework begins with initializing a random solution (S) and selecting a sequence of neighborhood operators N_k for $k = 1, \ldots, k_{max}$, each representing a specific neighborhood structure and collectively determining the order of neighborhood exploration in the search space. VNS then iterates until a termination criterion is met, systematically exploring each neighborhood in the sequence. For each $k = 1$ to $k_{\max}$, the following steps are performed: (a) a shaking procedure, designed to escape local optima, is applied to the current solution S to randomly generate a new solution S' from the kth neighborhood structure N_k of S, (b) a local search method (improvement procedure) is applied on S' to get a local optimum solution S'', and (c) an acceptance criterion is applied in which if S'' is superior to S, then S'' takes the place of S, and the procedure recenters its search from the first neighborhood N_1 (k=1); otherwise k is set to $k + 1$.

The idea behind VNS allows it to alternate between diversification (by exploring different neighborhood structures) and intensification (through local search), effectively guiding the search toward better solution quality within the search space.

Since the VNS framework is flexible, many variants of the Basic VNS has been developed. Notable variants include variable neighborhood descent (VND), where the change of neighborhoods is performed in a deterministic way; Reduced VNS (RVNS), which discards the local search step; General VNS in which more advanced improvement procedures are incorporated, unlike to Basic VNS, which uses a simple local search. For a comprehensive discussion of these variants, readers are referred to [32].

1.3 Scope of thesis

This thesis focuses on designing robust and effective metaheuristic techniques tailored to four $\mathcal{NP}$-hard combinatorial optimization problems. Metaheuristic techniques that are used are artificial bee colony (ABC) algorithm, iterated local search and variable neighborhood search, where ABC algorithm is a population-based metaheuristic technique, while ILS and VNS are single solution-based metaheuristic technique. The problems addressed in this thesis are: minimum edge-dilation k-center (MEDKC) problem, tree t-spanner problem (Tr_t-SP), single machine scheduling problem with periodic machine availability (PMASMSP), and min-degree constrained minimum spanning tree (MDMST) problem. Apart from being theoretically challenging, these problems find several real-world applications in routing schemes within computer networks; in various production environments where maintenance activities play a critical role in ensuring efficiency, safety, and improved productivity; distributed systems, communication networks, network design, robotics, and computational biology, particularly in building network synchronizers and reconstructing phylogenetic trees; and VLSI circuit design, data clustering, and telecommunications.

We have worked on artificial bee colony (ABC) algorithm and iterated local search (ILS) for the MEDKC problem; iterated local search (ILS) and Basic variable neighborhood search (VNS) for the Tr_t-SP; iterated local search (ILS) for the PMASMSP; and an improved variable neighborhood search (VNS^I) for the MDMST problem.

The research emphasizes the use of problem-specific knowledge in key components of metaheuristic techniques – such as encoding, solution initialization, and neighborhood operator – and appropriate operators to enhance performance. Where appropriate, suitable local search strategies or their extensions are integrated within the metaheuristic frameworks. The overarching goal is to ensure that these components operate cohesively to effectively explore the search space and deliver high-quality solutions within reasonable computational time.

This thesis is structured into six chapters, beginning with an introductory chapter. Chapters 2 through 5 detail the development and application of metaheuristic approaches – including artificial bee colony algorithm, iterated local search, and variable neighborhood search – on four selected $\mathcal{NP}$-hard combinatorial optimization problems. The final chapter offers concluding observations and highlights potential avenues for future research. A brief overview of each chapter's content is provided below.

Chapter 2 presents two metaheuristic techniques tailored to the MEDKC problem. The first one is artificial bee colony (ABC) algorithm, whereas the second one is iterated local search (ILS). In ABC algorithm, two neighborhood operators are employed in a mutual exclusive way specifically tailored to this problem. In ILS, a local search is proposed that effectively exploits the local region around the perturbed solution. Similar to two existing genetic algorithms (GA1 and GA2) [34], we have also used the same two set of available benchmark instances comprising of 91 instances. Our proposed approaches outperformed two existing genetic algorithms (GA1 and GA2) and a polynomial-time approximation algorithm (Konemann_AL) [35] across various performance metrics. In terms of best solution values, ABC algorithm showed improvements over GA1 on 18 instances and GA2 on 61 instances, while ILS showed improvements on 20 and 60 instances respectively. When average solution quality was considered, ABC algorithm surpassed GA1 on 68 instances and GA2 on 86, whereas ILS performed better on 65 and 69 instances respectively. Both our proposed approaches also consistently outperformed Konemann_AL on all 91 tested instances.

Chapter 3 presents two metaheuristic techniques – iterated local search (ILS) and Basic variable neighborhood search (VNS) – tailored to the tree t-spanner problem (Tr_t-SP).

The first proposed ILS relies on the problem-specific perturbation strategy involving two successive edge deletions and additions, a problem-specific local search improvement strategy, and an acceptance criterion that combines two strategies – where the second helps ILS escape deep local optima. The second proposed VNS systematically explores three neighborhood structures using problem-specific operators, and applies a local search improvement strategy. A restart criterion is incorporated to increase the chances of discovering new high-quality solutions. The performance of the proposed ILS and VNS is evaluated against the existing ABC algorithm [36] on available 48 benchmark Euclidean graph instances (with 50 to 200 vertices and varying edge densities). Results show that ILS outperforms existing ABC algorithm on 32 instances, is the same on 9, and worse on 7 in terms of best value (Best) over 10 runs; ILS is better on 26, the same on 1, and worse on 21 in terms of average solution quality (Avg) over 10 runs. Similarly, VNS outperforms existing ABC algorithm on 29 instances, is the same on 11, and worse on 8 in terms of best value (Best); ILS is better on 29, the same on 4, and worse on 15 in terms of average solution quality (Avg). Between VNS and ILS, VNS performs better on 13 instances, equal on 10, and worse on 25 instances in terms of best value. In terms of average solution quality, VNS is better on 23 instances, equal on 3, and worse on 22 instances compared to ILS. Furthermore, both ILS and VNS were found to be computationally faster than ABC algorithm on considered instances for the Tr_t-SP.

Chapter 4 presents an iterated local search (ILS) approach tailored to the PMASMSP. In the proposed ILS, the perturbation procedure is based on swapping of jobs in different periods, a series of local search strategies involving 1-0 insertion, 1-1 swap, 1-2 swap, and 2-1 swap, all focused on minimizing the processing time in the last period, and acceptance criterion follows two step procedures in which the second step, particularly, is used to prevent ILS to be trapped into deep local optima. The performance of the proposed ILS was evaluated against existing 19 approaches (15 constructive heuristic, 3 heuristics (NEW, NEW_FF and NEW_BF), and a MILP model) [37], and Genetic algorithm (GA) [38] on available benchmark instances. These instances are categorized into two class of benchmark instances called *LOW* and *MOD* instances generated by Paz and Jose [37]. Each class of instances has 14 categories in which the number of jobs vary within the range of {10, 20, 30, 40, 50, 60, 70, 80, 90, 100, 150, 200, 250, 300}. Each class of benchmark instances comprises of 700 instances. On

MOD instances, the performance of ILS is superior to all 19 approaches except three categories of instances with $n = 10, 20, 30$, where the performance of ILS is equivalent to that of MILP. When comparing with genetic algorithm, ILS performs better. In fact, ILS discovers new best results on 6 out of 14 categories of *MOD* instances. On LOW instances, the performance of ILS and GA are similar. On computational time part, ILS is computationally faster than GA.

Chapter 5 presents an improved variable neighborhood search (VNS^I), which is a general variable neighborhood search, tailored to the MDMST problem. In contrast to existing four VNS variants [39], VNS^I incorporates degree-aware edge selection within a multi-neighborhood framework comprising four structurally distinct neighborhoods (N_1–N_4). These are designed to prioritize edge removals that significantly influence the degree of vertex, thereby improving its ability to navigate the search space. Coupled with tailored two different local search strategies, this neighborhood design enhances both intensification and diversification capabilities. A restart criterion is incorporated to increase the chances of discovering new high-quality solutions. The performance of the proposed VNS^I was evaluated against the best-performing result among the existing four VNS variants under the label VNS^O [39], branch and cut approach [40], the best-performing result among the three variants (*gen-0*, *gen-1*, and *gen-2*) of genetic algorithm (min-GA*) [41], hybrid steady-state genetic algorithm (HSSGA) [42], iterated local search (ILS) and hybrid artificial bee colony algorithm (hABC) [43], and a population-based simulated annealing (PSA) [44] on available four standard benchmark datasets, namely CRD, SYM, ALM, and E250, covering 105 instances with varying graph sizes and structures. The proposed VNS^I consistently outperforms the existing VNS^O across CRD, SYM, and ALM data-sets, particularly in terms of solution quality. Overall, VNS^I dominates on ALM and E250 data-sets, and discovers 43 new best values across 105 instances – demonstrating its capability to search new values in the search space.

Chapter 6 concludes the thesis by summarizing its contributions. In addition, some directions are suggested for future research.

Chapter 2

Minimum Edge Dilation k-center Problem

2.1 Introduction

The MEDKC problem, proven to be a $\mathcal{NP}$-hard [35], can be outlined as follows: In a given edge-weighted, undirected and connected graph $G = (V, E, w)$ – with V denoting the vertex-set and E representing the edge-set, and w as the edge-weight function –, let $\Pi \subset V$ refer to a set of K center vertices and every other vertex $v \in V$ is allotted to only one vertex $\pi_v \in \Pi$, where $K > 0$ is a parameter. The objective of the MEDKC problem is to identify a set Π of K center vertices and assign each remaining vertex to a unique vertex in Π in a manner that minimizes the stretch of the solution $(\Pi, \pi_{v_{v \in V}})$. The stretch of a solution $(\Pi, \{\pi_v\}_{v \in V}) = \max\limits_{(u,v) \in V \times V} \{\frac{d_\pi(u,v)}{d_w(u,v)}\}$ is described as the maximum stretch among all pairs of vertices in G.

The stretch for a pair of vertices $(u, v \in V)$ is the ratio of the center distance $(d_\pi(u, v))$ to the shortest distance $(d_w(u, v))$ in G, represented as $\frac{d_\pi(u,v)}{d_w(u,v)}$, where $d_\pi(u, v)$ with respect to Π is defined as

$$d_\pi(u, v) = d_w(u, \pi_u) + d_w(\pi_u, \pi_y) + d_w(\pi_y, v) \tag{2.1}$$

Here, $d_w(u, v)$ signifies the shortest path between u and v.

The optimal solution for a given G, denoted as $Opt(G) = \min(\Pi, \{\pi_v\}_{v \in V})$ over all possible subsets Π and assignment $\{\pi_v\}_{v \in V}$.

In the literature, the MEDKC problem exhibits connections to various problems [45]. Some of these related problems include:

- □ p-center problem aims to seek a set of p vertices $\in V$ such that the maximum distance from any vertex to its closest center is as minimum as possible [45, 46].
- □ The capacitated K-center problem allows each center to attend only a certain number of vertices [47]
- □ The fault tolerant K-center problem in which each selected center must belong to a set of $\alpha \leq K$ centers close to it [48].
- □ The p-next center problem that aims to minimize the total sum of the distance from the farthest vertex to its nearest center and the distance between this center to its nearest alternative center [49].
- □ The mixed K-center problem in which q centers must be in the set of vertices, and the remaining vertices can be anywhere ($q < K$) [50].

Könemann *et al.* [35] introduced the MEDKC problem, inspired by a practical application in routing schemes within computer networks. In such schemes, each source node has the capability to route messages to a destination node. A host (node) stores information about routing paths to every other host in its routing table. To ensure shortest-path routing for the entire network, each node requires $\mathcal{O}(n)$ entries in its routing table to retain information about the shortest-path routing to every other node.

Considering computer networks in modern times comprising of millions of nodes poses a challenge for each node to store information on shortest-path routing for other nodes because of memory limitations. Attempts have been made to develop an efficient routing scheme, prioritizing considerations for memory constraints. Researchers focus on two main objectives: minimize the size of routing tables and/or minimize the stretch of a routing scheme. Modern routing protocols such as OSPF [51] enable the division of a network into areas, where each node retains information about paths within the same area. Routing between nodes in distinct areas is accomplished through a backbone network of area border routers connecting these areas. This results in a decrease in the routing table size for each node, with the anticipation that the overall network stretch remains acceptable.

The objective of the MEDKC problem is based on this principle that K center nodes in computer networks are exactly the area border routers, while each remaining node in the network is assigned to a router. The task is to construct a compact routing scheme that minimizes the maximal stretch of the entire network.

2.2 Literature review

The literature has seen numerous papers addressing compact routing schemes, with a primary emphasis on balancing the trade-off between the size of routing tables and the stretch. In the initial stages, Peleg and Upfal [52] pioneered compact routing schemes tailored for undirected graphs. Subsequently, a routing scheme [53][54] was proposed for weighted graphs. Thorup and Zwick [55] introduced a labeled compact routing scheme with a stretch of $4K$-5 using a routing table of size $\mathcal{O}(\frac{1}{k})$-bit by each vertex. Chechik [56] enhanced the outcome of [55]. Eilam *et al.* [57] employed the pivot interval routing (PIR) scheme, achieving a stretch factor of at most 5 and an average stretch factor of at most 3. Each vertex utilized a routing table of size $\mathcal{O}(\sqrt{n}\log^{3/2} n)$ bits. Cowen [58] demonstrated that if $\mathcal{O}(n/2/3\log^{4/3} n)$ space is allowed at each vertex in a given weighted undirected network (G), a solution can be found where the routing path between any pair of vertices $((u, v \in V))$ is at most 3 times as long as the shortest u, v-path in G. Krioukov *et al.* [59] established that using sub-linear sized routing tables, a stretch of less than 3 is achievable. Enachescu *et al.* [60] applied Internet-like graphs and proposed a routing scheme with a stretch value of less than 3 with high probability while retaining $o(n)$ memory. Later, Abraham *et al.* [61] presented a routing scheme for a given weighted undirected network achieving a stretch of 3 using a $\mathcal{O}(\sqrt{n})$ space at each vertex. Roditty and Tov [62] pintroduced a routing scheme that surpasses previous results achieving a stretch of 5+ϵ using $\mathcal{O}(\frac{1}{\epsilon}n^{1/3}\log D)$ memory at each node, the label of each node of $\mathcal{O}(\log n)$ size, and $\mathcal{O}(\frac{1}{\epsilon}\log D)$-bit headers.

Based on the literature discussed above, it is evident that there was no existing solution approach for the MEDKC problem applicable to practical scenarios. Moreover, the methods developed for creating a compact routing scheme cannot be directly employed, given the distinct characteristics of the two problems. Compact routing scheme problems are concerned with balancing the trade-off between stretch and the size of routing tables. In contrast, the MEDKC problem, where memory that is used to store distance information is not an issue, aim to find a set of K central nodes and to arrange the paths over them in a manner that minimizes the maximum stretch [35].

Könemann *et al.* [35] proved that the MEDKC problem is a $\mathcal{NP}$-hard problem. They also introduced a polynomial time approximation algorithm (referred to as *Konemann_AL*), which is at most 4·Opt + 3, where Opt represents the optimal stretch value. The study on the results for the MEDKC problem [35] is of theoretical importance, as their proposed approximation algorithm provides only an upper bound for any solution. Given its status as an $\mathcal{NP}$-hard problem [35], employing metaheuristic techniques becomes a favorable approach to obtain high-quality solutions within a reasonable computational timeframe. Readers are suggested to study a tutorial on metaheuristic techniques [63]. To the best of our knowledge, among metaheuristic techniques only Matic *et al.* [34] proposed two variants (GA1 and GA2) of genetic algorithm which are evolutionary algorithms. In both variants of GA, authors used binary coding and used modified crossover and mutation operators, particularly for each GA variant, in order to maintain feasibility of solutions throughout the optimization process. Based on the literature survey, it is clear that the MEDKC problem is an under-studied problem in terms of metaheuristic techniques.

2.3 Our Contribution

This chapter presents two metaheuristic techniques tailored to the MEDKC problem. The first approach is based on swarm intelligence-based artificial bee colony (ABC) algorithm while the second is iterated local search (ILS). The motivation for applying ABC algorithm and ILS is that neither has been applied to this problem before. The proposed ABC algorithm uses two adaptable neighborhood operators tailored to the MEDKC problem, while in ILS framework, we propose a local search that effectively exploits the local region around the perturbed solution. On available benchmark instances, computational results indicate that the proposed both approaches overall performs superior to the existing two variants (GA1 and GA2) of genetic algorithm in terms of solution quality and computational time and they also dominates an existing polynomial time approximation algorithm (*Konemann_AL*) in terms of finding solution quality.

The remainder of the chapter is summarized as follows: Section 2.4 and 2.5 present ABC algorithm and ILS tailored to the MEDKC problem, respectively; and Section 2.6 discusses the computational results of our proposed approches in comparison to the existing approaches.

2.4 ABC algorithm for the MEDKC problem

This section describes our proposed ABC algorithm for the MEDKC problem. In the beginning of ABC algorithm, precompute the shortest paths between all pairs of vertices of a given input graph (G).

The salient features of ABC algorithm for the MEDKC problem are discussed in the following subsections:

2.4.1 Solution encoding

Each feasible solution is denoted as a set of K vertices acting as centers. Vertices that are not included in the K center vertices of the solution are classified as non-center vertices.

2.4.2 Initial solution generation

Algorithm 4: Pseudo-code of generation of an initial solution E_i

$E_i \leftarrow \emptyset$;
$U \leftarrow \emptyset$;
$U \leftarrow U \cup \{i\} \ \forall i \in V$;
for $i \leftarrow 1$ **to** K **do**
    Select a random vertex r as a center vertex from U;
    $E_i \leftarrow E_i \cup \{r\}$;
    $U \leftarrow U \setminus \{r\}$;

Each initial solution, denoted as E_i, within the employed bee population, is randomly generated, where the size of the employed bee population is E_{pop}. The initial solution E_i, designed to contain K center vertices, begins as an empty set. Initially, all vertices are assigned to a set called U. In the subsequent steps, the procedure, at each iteration, randomly selects a vertex (denoted as r) from U and assigns it as the center vertex for E_i. The set U is then updated by removing the selected vertex r. This iterative process continues until K vertices are assigned as center vertices to E_i. Vertices not included in E_i remain in set U and are considered non-center vertices. Algorithm 6 provides the pseudo-code for the procedure outlining the generation of the initial solution E_i.

2.4.3 Fitness computation

Upon the completion of constructing a feasible solution (denoted as S), the fitness, represented by the stretch of S, is computed by determining the maximum stretch among all pairs of vertices in G. This is defined as follows:

$$\max_{(u,v)\in V\times V}\left\{\frac{d_\pi(u,v)}{d_w(u,v)}\right\}$$

Readers are suggested to refer to the introduction section (Section 2.1) for more details of the stretch of a feasible solution in the context of the MEDKC problem.

2.4.4 Probability of selecting a solution

Each onlooker bee uses a binary tournament selection method to select a solution from the employed bee population. This selection mechanism involves randomly selecting two different solutions from the employed bee population. The preferred solution, based on the fitness defined in Section 2.4.3, is chosen with a probability P_{bt}, while the less favorable one is chosen with a probability of 1-P_{bt}. This method prioritizes better solutions over inferior ones, encouraging more onlooker bees to opt for higher-quality solutions. Additionally, this selection method is also employed to choose a distinct solution in the first neighborhood operator (refer to Section 2.4.5.1) when determining a new neighboring solution.

2.4.5 Neighborhood operators

Neighborhood operators in ABC algorithm play a significant role in finding high-quality solutions for $\mathcal{NP}$-hard combinatorial optimization problems [64, 65, 66]. For the MEDKC problem, we apply two neighborhood operators in a mutually exclusive way to determine a new neighboring solution (say E_i') within the vicinity of the current employed bee solution (say E_i) of employed bee population. The first neighborhood operator (*Nbr_cv*) is based on this fact that if a center vertex is present in a solution with high fitness, the chance of that center vertex being present in many other solutions of the employed bee population is high, whereas the second neighborhood (*Nbr_swp*) is based on the swapping of center vert(ex/ices) with non-center vert(ex/ices) in a mixed strategies of greedy and random approach. The details of each neighborhood are outlined as follows:

2.4.5.1 First neighborhood operator (Nbr_cv)

The first neighborhood operator (*Nbr_cv*) first creates a copy (say E_i') of the current employed bee solution (say E_i), and then, it selects a solution (say E_j), different from E_i, by using binary tournament selection method (see Section 2.4.4 in detail). A check is performed whether E_i and E_j are identical. If both are different, then a certain number (say *nocv_1*) of center vertices of E_j is selected randomly, where $nocv\_1 = max(1, (int)(Para1 \times cnt))$. Here, *cnt* in E_j is the number of those center vertices which are different from the center vertices in E_i. *Para1* is a parameter whose value is determined empirically. All these selected center vertices from E_j are assigned to E_i', making solution E_i' infeasible. To make E_i' feasible, the same number (*nocv*) of center vertices which are originally part of E_i is removed from E_i'. This way makes E_i' feasible, resulting a new feasible neighboring solution E_i'.

It is important to emphasize that if both E_i and E_j are identical, the continuation of the first neighborhood operator (*Nbr_cv*) is halted, a scenario referred to as a *collision* [20]. Such duplication of two solutions within the employed bee population poses an obstacle to effectively searching for high-quality solutions in the search space. To handle this issue, the employed bee associated with E_i is designated as a scout. The scout bee then explores and identifies a new solution (refer to Section 2.4.6). This approach aids in eliminating a duplicated solution from the employed bee population, promoting greater diversity within the population. However, in the event of a collision during the onlooker bee phase, the first neighborhood operator (*Nbr_cv*) is similarly discontinued for the associated solution, and a very large fitness value is assigned to that particular solution associated by its onlooker bee.

2.4.5.2 Second neighborhood operator (Nbr_swp)

The second neighborhood operator (*Nbr_swp*) is based on the swapping of center vert(ex/ices) with non-center vert(ex/ices) in a mixed strategies of greedy and random approach. *Nbr_swp* initiates with creating a copy (say E_i') of the current employed solution E_i. Subsequently, *Nbr_swp* randomly selects a certain number (say *nocv_2*) of center vertices from E_j, where $nocv\_2 = max(1, (int)(Para2 \times K))$. Here K is the number of center vertices. *Para2* is a parameter whose value is to be determined empirically. Each selected random center vertex (say v_c) of E_i' is replaced with a candidate non-center vertex (v_{nc}) of E_i' in either two ways. With probability (*Para*3), *Nbr_swp* applies a greedy approach in which a candidate non-center vertex is selected from a subset of total number of non-center vertices of E_i' (say $no\_ncv = (int)(val \times (V\text{-}K))$, where *val* is set to 0.1 based on our preliminary experiments in order to balance a trade-off between running time and solution quality) based on having minimum fitness. Otherwise, *Nbr_swp* applies a random approach wherein a candidate non-center vertex is selected randomly from the total number of non-center vertices of E_i' with probability (1-*Para*3).

2.4.6 Scout bee phase

During the scout bee phase, if an employed bee solution ceases to show improvement for a specific number of generations (referred to as *limit*), the employed bee promptly abandons its non-improving solution, transitioning its status to that of a scout bee. The scout bee then seeks a new solution through a prescribed procedure (refer to Section 2.4.5.2). Once the process for finding a new solution concludes, the scout bee reverts to the state of an employed bee. The parameter *limit* to be determined empirically holds substantial importance in the ABC algorithm. In addition, a collision (as described in Section 2.4.5.1) can also prompt the transformation of an employed bee into a scout. It is crucial to note that the ABC algorithm does not impose an upper limit on the number of scouts in a single generation. A generation may witness multiple scouts if the conditions mentioned earlier (*limit* and *collision* conditions) are met; otherwise, no scouts are encountered.

Algorithm 8 outlines the pseudo-code of ABC algorithm for the MEDKC problem in which the procedure $BTSM(E_1, E_2, \ldots, E_{E_{pop}})$ is a binary tournament selection method

Algorithm 5: The pseudo-code of ABC algorithm for the MEDKC problem

```
Generate initial solutions < E_1, E_2, ..., E_{E_pop} >;
Best ← best solution in < E_1, E_2, ..., E_{E_pop} >;
while Termination criterion is not reached do
    for i ← 1 to E_pop do
        if u01 < P_nbr then
            In_i ← BTSM(E_1, E_2, ..., E_{E_pop});
            if E_{In_i} is identical to E_i then
                goto Jump_1;
            E'_i ← Nbr_cv(E_i, E_{In_i});
        else
            E'_i ← Nbr_swp(E_i);
        if F(E'_i) < F(E_i) then
            E_i ← E'_i;
            if F(Best) > F(E_i) then
                Best ← E_i;
        else if There is no improvement in E_i for a certain number of generations (i.e. limit)
         then
            Jump_1;
            Scout bee phase;
            if F(Best) > F(E_i) then
                Best ← E_i;
    for i ← 1 to On_pop do
        ind_i ← BTSM(E_1, E_2, ..., E_{E_pop});
        if u01 < P_nbr then
            In_j ← BTSM(E_1, E_2, ..., E_{E_pop});
            if E_{In_j} is identical to E_{ind_i} then
                F(O_i) ← ∞;
            else
                O_i ← Nbr_cv(E_{ind_i}, E_{In_j});
        else
            O_i ← Nbr_swp(E_{ind_i});
    for i ← 1 to On_pop do
        if F(O_i) < F(E_{ind_i}) then
            E_{ind_i} ← O_i;
        if F(Best) > F(E_{ind_i}) then
            Best ← E_{ind_i};
```

that provides an index of a solution in $< E_1, E_2, \ldots, E_{E_{pop}} >$; $Nbr\_cv(X_1, X_2)$ is the first neighborhood operator that provides a new neighboring solution (say X'); and $Nbr\_swp(X)$ is the second neighborhood operator that provides a new neighboring solution (say X').

2.5 ILS for the MEDKC problem

This section discusses ILS approach for the MEDKC problem. Before starting ILS approach for this problem, shortest paths between all pairs of vertices of a given instance G of this problem are precomputed.

The following subsections discuss the main components of ILS:

2.5.1 Solution encoding

ILS follows the same solution encoding which is used in ABC algorithm (see Section 2.4.1) for the MEDKC problem.

2.5.2 Generation of an initial solution

The initial solution of the ILS framework for the MEDKC problem is generated by a randomized initialization procedure. This randomized procedure starts with an empty solution (say S). All vertices are assigned to a set (say U). The procedure, then at each iteration, picks up a vertex (say r) randomly from U and adds this vertex r to S as a center vertex. Update U by removing r from U, i.e., $U \leftarrow U \setminus r$. This iteration is continued again and again until K number of vertices are added to S as center vertices. Vertices which are not in S are in U and are known as non-center vertices. Algorithm 6 presents the pseudo-code of this randomized initialization procedure for generating an initial solution S.

2.5.3 Computation of the fitness of a solution

Once a feasible solution (say S) is determined, the fitness (stretch) of S (i.e. $F(S)$) is computed through finding the maximum stretch over all pairs of vertices in G which is defined as follows:

Algorithm 6: Pseudo-code of the randomized initialization procedure for generating an initial solution (S)

```
S ← ∅;
U ← ∅;
U ← U ∪ i  ∀i ∈ V;
for i ← 1 to K do
    Select a random vertex (say r) as a center vertex from U;
    S ← S ∪ {r};
    U ← U \ {r};
```

$$\max_{(u,v)\in V\times V}\left\{\frac{d_\pi(u,v)}{d_w(u,v)}\right\}$$

Readers are suggested to read the introduction section (Section 2.1) for more details of the stretch of a feasible solution for the MEDKC problem.

2.5.4 Perturbation

The role of perturbation procedure (say *Perturb*(S)) is to move the current solution (S) from its current position to another position in the search space. In this perturbation procedure, the procedure initially creates a copy (say S') of S. The procedure, then, selects a random center vertex (say *nrcv*=1) from S'. This selected center vertex is swapped with a non-center vertex of S' which is also selected randomly. This perturbed solution becomes a new starting point (solution) in the search space for the local search.

2.5.5 Local search

The design of the local search strategy becomes important for the success of ILS for the problem under consideration, as it has to exploit the local region of the perturbed solution (S') that leads to a local optimum solution. Our proposed local search procedure (*LS* (S')) consists of two nested loops. In each iteration of the outer loop, the local search procedure selects a center vertex (say v_c) randomly, then the inner loop starts. In each iteration of the inner loop, a non-center vertex (say u_i) is selected randomly, and a swap is performed between v_c and u_i. If this swap leads to an improved solution, then this non-center vertex and the fitness of the resultant solution are memorized. The

Algorithm 7: Local search procedure for the MEDKC problem

Input: A perturbed solution (S')

Output: A local optimum solution (S'')

$Y = (int)(K \times Para_1)$;`//`$Para_1$ `is a parameter`
if $Y > |U|/2$ **then**
    $Y \leftarrow |U|/2$; `//`U `is a set of non-center vertices of` S'
$S'' \leftarrow S'$;
$z \leftarrow 0$;
$Improve \leftarrow 0$;
$Count \leftarrow 4$;
while $z < K$ *and* $Improve < Count$ **do**
    Select a center vertex (v_c) randomly from S'';
    $Min \leftarrow F[S']$;
    $mem \leftarrow$ -1;
    for $i \leftarrow 1$ **to** Y **do**
        Select a non-center vertex (say u_i) randomly from U of S'';
        Swap v_c with u_i that leads to a new solution;
        if $Min >$ *the fitness of this new solution* **then**
            $Min \leftarrow$ the fitness of this new solution;
            $mem \leftarrow u_i$;
        Restore the solution after swapping u_i with v_c;
    if *mem* > *-1* **then**
        $F[S'] \leftarrow Min$;
        Swap center vertex v_c with non-center vertex *mem* that leads to a new improved solution (S'');
        $Improve \leftarrow Improve + 1$;
    $z \leftarrow z + 1$;
return S'';

solution is restored after swapping u_i with v_c. The inner loop is continued for a certain number (say Y) of iterations. When the inner loop stops iterating, then it is examined whether a non-center vertex whose swap with v_c leads to the maximum improvement in terms of fitness exists (has been memorized) or not. If it exists, this non-center vertex is finally swapped with v_c. The outer loop of the local search procedure stops iterating when one of two conditions becomes false, where the first condition becomes false when the outer loop iterates more than K iteration and the second condition becomes false when the value of $Improve$ is greater than the value of $Count$. It is to be noted that $Count$ is a paramter. The local search procedure leads to a new local optimum solution S''. Algorithm 7 presents the pseudo-code of the local search procedure. In Algorithm 7, $Para_1$ is a parameter which is used to determine the value of Y.

Although the local search procedure is a computationally expensive procedure, the values of Y and $Count$ have been used in the local search procedure so that a trade-off between solutions of high quality and their computing time is maintained.

2.5.6 Acceptance Criterion

A sequence of the perturbation procedure and local search procedure leads to a possible change between a current solution (S) to a locally optimum solution (S''). The acceptance criterion procedure ($Acceptance\_Criterion(S, S'')$) then helps in taking decision whether the currently evaluated new local optimum solution is a promising initial point for the next iteration (generation) in the search space or not. The procedure ($Acceptance\_Criterion(S, S'')$) plays a decisive role in controlling the balance between exploitation and exploration during the search.

Algorithm 8 presents the pseudo-code of ILS for the MEDKC problem.

It is important to highlight that the most time-consuming part of ABC algorithm and ILS is the computation of the fitness for each generated solution. This process involves computing shortest paths between all pairs of vertices in a given input $G = (V, E, w)$, incurring a time complexity of $\mathcal{O}(V^3)$. Additionally, it encompasses (i) the computation of center distance for each pair of vertices, with a complexity of $\mathcal{O}((V\text{-}K) \times K)$, and (ii) the computation of stretch for each pair of vertices, with a complexity of $\mathcal{O}((V\text{-}K)^2)$ (refer to the first two paragraphs of Section 2.1). To alleviate this computational load, the approach precomputes the shortest paths between all pairs of vertices in a given input $G = (V, E, w)$ once. Consequently, the fitness function procedure for each generated solution utilizes this precomputed information of shortest paths between all pairs of vertices, reducing the running time from $\mathcal{O}(V^3)$ to $\mathcal{O}((V\text{-}K)^2)$. Another time-

Algorithm 8: The pseudo-code of ILS for the MEDKC problem

```
Generate an initial solution (S) with the help of a randomized initialization
  procedure;
Best ← S;
while the stopping criterion is not met do
    S′ ← Perturb(S); //Apply Perturbation procedure on S
    S″ ← LS(S′); //Apply Local Search on S′
    //Acceptance_Criterion(S, S″)
    if F(S″) < F(S) then
        S ← S″;
        if F(Best) > F(S) then
            Best ← S;
```

consuming part of ABC algorithm involves a greedy approach applied by Nbr_swp with probability (*Para*3). In this procedure, a selected random center vertex is tried for swap with a subset of total number of non-center vertices (*no_ncv*) one-by-one, and the swap that results in the minimum fitness is chosen. The running time for this greedy approach is $\mathcal{O}(no\_ncv \times (V\text{-}K)^2)$, where $\mathcal{O}((V\text{-}K)^2)$ is the running time for computing the fitness of a newly generated solution after a swap.

2.6 Computational results

C language is used to code our approches ABC algorithm and ILS for the MEDKC. Experiments were conducted on a Linux system equipped with an Intel Core *i*5-4570 CPU @ 3.2 GHz × 4 processor and 4 GB of RAM using a single thread. Similar to [34], the performance of our proposed approaches are tested on the same set of benchmark instances. For each instance, both ABC algorithm and ILS were allowed to execute 20 independent runs. The termination criterion for ABC algorithm allows to execute for either at least 150 generations or until the best-so-far obtained solution does not improve after ($K \times 30$) generations. For termination criterion of ILS, it is allowed to execute either at least *Gen_val* generations or until the best-so-far obtained solution does not improve after 50 generations, where $Gen\_val = 100$ if $K < 10$ && $|V| < 400$, otherwise *Gen_val* = 150.

The subsequent subsections discuss about instances, parameter settings and computational results of our approaches in comparison to existing approaches, including a polynomial-time approximation algorithm (*Konemann_Kon*) [35] and two variants of genetic algorithms, namely GA1 and GA2 [34].

2.6.1 Benchmark Instances

Similar to GA1 and GA2 [34], we have also used the same two set of instances available in the public domain. The first set pertains to the BX instances, comprising 7 graphs that are 4-regular and possess varying vertex sizes from the set {25, 50, 75, 100, 200, 400, 1000} [67]. The second set corresponds to the BB instances, including 7 graphs: 5 edge-weighted graphs of grid with sizes ranging from $< 15 \times 15, 33 \times 33, 45 \times 5, 50 \times 50$ and $100 \times 10 >$, and two large Steiner tree benchmark instances, namely steinc5 and steind5, each having 500 vertices and 1000 vertices respectively [68]. Each instance is examined for different values of K with in the set $\in \{2, 3, 4, 5, 10, 15, 20\}$. It is important to note that for K=1, the MEDKC problem is not $\mathcal{NP}$-hard [34].

2.6.2 Parameter Tuning

Given its stochastic nature, both of our proposed approaches (ABC algorithm and ILS) for the MEDKC problem rely on a set of parameters, each of which plays a critical role in their effectiveness. The proper configuration of these parameters is essential for achieving high-quality solutions. Although tuning these parameters can be challenging, it is feasible to approximate suitable value of each parameter value in a way that helps ABC algorithm and ILS in finding solutions of high quality.

To determine the value of each parameter used in ABC algorithm, we examined it on ten instances, i.e. bb15×15_1 (k=20), bb15×15_1 (k=4), g25-4-01 (k=10), g50-4-01 (k=3), g100-4-01 (k=15), g200-4-01 (k=15), bb45×5_1 (k=20), bb15×15_1 (k=15), g50-4-01 (k=15), and g50-4-01 (k=10). Drawing from existing literature and our prior experience with ABC algorithm, we considered a range of potential values for each parameter (refer to Table 2.1). Initial experiments indicated that ABC algorithm for the MEDKC problem exhibited the best overall performance across the ten instances under consideration when E_{pop}=10, On_{pop}=30, *limit*=25, P_{bt}=0.85, *P_nbr*=0.7, *Para*1=0.1,

*Para*2=0.1 and *Para*3=0.5.

Similarly, to determine the value of each parameter used in ILS, we examined it on four instances (see Table 2.2). A range of potential values of each parameter of ILS (see Table 2.2) is considered in order to achieve a trade-off solution quality and computational time. The results (Table 2.2) indicate that ILS performs best overall when the parameter values are set to Count = 4 and Para1 = 25. The best-performing values of each parameter, along with the corresponding solution quality, are highlighted in bold in Table 2.1 and Table 2.2.

Table 2.1: Influence of parameter setting for ABC algorithm on the solution quality

Parameter	Value	bb15x15_1 (k=20)		bb15x15_1 (k=4)		g25-4-01 (k=10)		g50-4-01 (k=3)		g100-4-01 (k=15)		g200-4-01 (k=15)		bb45x5_1 (k=20)		bb15x15_1 (k=15)		g50-4-01 (k=15)		g50-4-01 (k=10)	
		Best	Avg	Best	Avg	Best	Avg	Best	Avg	Best	Avg	Best	Avg	Best	Avg	Best	Avg	Best	Avg	Best	Avg
E_{pop}	5	18.33	**19.11**	13400	139	2.16	2.32	11.73	11.79	7.38	7.7	10.69	10.99	17.67	18.25	24.71	25.59	4.32	4.35	5.46	5.48
	10	**18.33**	19.71	**134.00**	138	**2.16**	**2.16**	**11.73**	**11.73**	**7.38**	**7.61**	**10.62**	11.01	**17.67**	18.54	**24.71**	**25.16**	**4.32**	**4.32**	**5.46**	**5.46**
	20	19	20.14	134	135.43	2.16	2.17	11.73	11.73	7.38	7.68	10.92	11.24	18	19.01	24.71	25.68	4.32	4.32	5.46	5.47
	30	19.5	20.41	134	**134.12**	2.16	2.16	11.73	11.73	7.38	7.76	11	11.27	18.33	19.44	24.71	26.11	4.32	4.32	5.46	5.46
On_{pop}	10	19.5	20.82	134	140	2.16	2.19	11.73	11.77	7.38	7.88	10.69	11.53	19	20.48	25	27.14	4.32	4.4	5.46	5.57
	20	18.33	20.12	134	141.02	2.16	2.18	11.73	11.78	7.38	7.73	10.79	11.14	18	19.52	24.71	25.73	4.32	4.35	5.46	5.5
	30	**18.33**	**19.71**	**134.00**	**134.88**	**2.16**	**2.16**	11.73	**11.73**	**7.38**	**7.61**	**10.62**	11.01	**17.67**	18.54	**24.71**	**25.16**	**4.32**	**4.32**	**5.46**	**5.46**
limit	10	18.33	20.51	134	**135.68**	2.16	2.17	**11.73**	11.73	7.38	7.75	11	11.15	17.67	19.93	24.71	26.36	4.32	4.33	5.46	5.48
	20	19.5	20.82	134	140.3	2.16	2.17	11.73	11.74	7.38	7.68	10.62	11.08	17.67	18.86	24.71	25.44	4.32	4.34	5.46	5.51
	25	**18.33**	19.71	**134.00**	138.88	**2.16**	**2.16**	11.73	**11.73**	**7.38**	**7.61**	**10.62**	11.01	**17.67**	18.54	**24.71**	**25.16**	**4.32**	**4.32**	**5.46**	**5.46**
	50	18.33	**19.27**	134	139.6	2.16	2.21	**11.73**	11.73	7.38	7.62	10.69	10.96	17.67	18.52	24.71	25.41	4.32	4.34	5.46	5.47
P_{bt}	0.8	18.33	**19.27**	134	139.6	2.16	2.17	11.73	11.73	7.38	**7.57**	10.69	11.03	17.67	18.47	24.71	25.16	4.32	4.33	5.46	5.46
	0.85	**18.33**	19.71	**134.00**	**138.88**	**2.16**	**2.16**	11.73	**11.73**	**7.38**	7.61	**10.62**	11.01	**17.67**	18.54	**24.71**	**25.16**	**4.32**	**4.32**	**5.46**	**5.46**
	0.9	18.33	19.88	134	139.6	2.16	2.18	**11.73**	11.83	7.38	7.63	10.69	11.03	18	18.73	24.71	25.87	4.32	4.33	5.46	5.46
P_nbr	0.6	18.33	**19.50**	134	**135.40**	2.16	2.18	11.73	11.73	7.38	**7.56**	10.69	**10.95**	17.67	18.62	24.71	25.23	4.32	4.33	5.46	5.46
	0.7	**18.33**	19.71	**134.00**	138.88	**2.16**	**2.16**	11.73	**11.73**	**7.38**	7.61	**10.62**	11.01	**17.67**	18.54	**24.71**	**25.16**	**4.32**	**4.32**	**5.46**	**5.46**
	0.8	18.33	19.88	134	141.27	2.16	2.22	**11.73**	11.74	7.38	7.7	10.62	11.09	17.67	18.62	24.71	25.87	4.32	4.33	5.46	5.49
Para1	0.05	18.57	**19.68**	134	138.88	2.16	2.16	11.73	11.74	7.38	7.61	10.62	11.01	17.67	18.55	24.71	25.16	4.32	4.32	5.46	5.46
	0.1	**18.33**	19.71	**134.00**	**138.88**	**2.16**	**2.16**	11.73	**11.73**	**7.38**	**7.61**	**10.62**	11.01	**17.67**	18.54	**24.71**	**25.16**	**4.32**	**4.32**	**5.46**	**5.46**
	0.2	18.33	19.79	134	138.88	2.16	2.2	**11.73**	11.73	7.38	7.75	10.62	11	17.67	19.17	24.71	25.18	4.32	4.34	5.46	5.46
Para2	0.05	18.33	**18.84**	134	138.88	2.16	2.16	11.73	11.73	7.38	7.61	10.62	11.01	17.67	**17.95**	24.71	25.16	4.32	4.32	5.46	5.46
	0.1	**18.33**	19.71	**134.00**	**138.88**	**2.16**	**2.16**	11.73	**11.73**	**7.38**	**7.61**	**10.62**	11.01	**17.67**	18.54	**24.71**	**25.16**	**4.32**	**4.32**	**5.46**	**5.46**
	0.2	20.00	20.86	134	141.27	2.16	2.18	**11.73**	11.73	7.5	7.93	11.18	12.3	18.08	20.11	24.71	25.87	4.32	4.37	5.46	5.51
Para3	0.4	18.33	19.71	134	138.88	2.16	2.16	11.73	11.73	7.38	7.61	10.62	11.01	17.67	18.54	24.71	25.16	4.32	4.32	5.46	5.46
	0.5	**18.33**	**19.71**	**134.00**	**138.88**	**2.16**	**2.16**	11.73	**11.73**	**7.38**	**7.61**	**10.62**	11.01	**17.67**	18.54	**24.71**	**25.16**	**4.32**	**4.32**	**5.46**	**5.46**
	0.6	18.33	19.71	134	138.88	2.16	2.16	**11.73**	11.73	7.38	7.61	10.62	11.01	17.67	18.54	24.71	25.16	4.32	4.32	5.46	5.46

Table 2.2: Influence of parameter setting for ILS on the solution quality

Instance	K	*Count*=2		*Count*=3		*Count*=4		$Para_1$=10		$Para_1$=20		$Para_1$=25	
		Best	Avg	Best	Avg	Best	Avg	Best	Avg	Best	Avg	Best	Avg
g400-4-01	20	16.200001	18.622717	16.200001	17.899490	**15.560000**	**17.169756**	**15.560000**	**17.169756**	**15.560000**	**17.169756**	**15.560000**	**17.169756**
g1000-4-01	10	54.000000	59.908337	**53.000000**	**56.463329**	**53.000000**	57.008339	53.500000	56.833336	**53.000000**	57.011665	**53.000000**	**57.008339**
g1000-4-01	15	43.666668	47.600002	43.666668	46.806667	**43.000000**	**45.033337**	43.500000	45.709446	**43.000000**	45.986664	**43.000000**	**45.033337**
100 x 10	10	**375.000000**	394.000000	**375.000000**	390.700012	**375.000000**	**389.899994**	**375.000000**	393.299988	**375.000000**	391.399994	**375.000000**	**389.899994**

2.6.3 Comparison of results of ABC algorithm and ILS against state-of-the-art approaches for the MEDKC problem

This section discusses a comparative study on the computational results of ABC algorithm and ILS with the computational results of state- of-the-art approaches (i.e. *Konemann_AL*) [35], and genetic algorithms (GA1, and GA2) [34] on the BX and BB instances.

Tables 2.3-2.4 presents the results of *Konemann_AL*, GA1, GA2, ABC algorithm and ILS on the 49 BX instances and 42 BB instances respectively. In these tables 2.3-2.4, *Instance* column indicates the name of the instance; $|V|$ column represents the number of vertices in G; the column *Value* represents the value of solution returned by *Konemann_AL*; the next two three columns, i.e., *Best*, *Avg*, $ATTB$ and $ATET$, respectively, represent the best-so-far value, average value, average time to reach the best solution (in seconds), and average total execution time (in seconds) over 20 runs obtained by GA1 and GA2; the next four columns, i.e., *Best*, *Avg*, SD, and $ATTB$, respectively, represent the best-so-far value, average value, standard deviation, average time to reach the best solution (in seconds), and average total execution time over 20 runs obtained by ABC algorithm; and the last four columns, i.e., *Best*, *Avg*, SD, and $ATTB$, respectively, represent the best-so-far value, average value, standard deviation, average time to reach the best solution (in seconds), and average total execution time over 20 runs obtained by ILS. The best *Avg* results are highlighted in bold in both tables 2.3-2.4. The code for the polynomial-time approximation algorithm (*Konemann_AL*) [35] for the MEDKC problem is available to us, we executed this code on considered instances using the same computer configuration used for ABC algorithm and ILS then reported the results of *Konemann_AL* in tables 2.3-2.4.

Based on the overall results across 91 instances reported in Tables 2.3-2.4, the following observations can be made:

- ABC vs. GA1: In terms of *Best* values, ABC performs better on 18 instances, equal on 60, and worse on 13. For *Avg* values, ABC is better on 68, equal on 14, and worse on 9 instances.
- ABC vs. GA2: For *Best* values, ABC outperforms GA2 on 61 instances, is equal on 26, and worse on 4. In terms of *Avg* values, ABC is better on 86 instances,

equal on 2, and worse on 3.

- ◻ ILS vs. GA1: ILS performs better in terms of *Best* values on 20 instances, equal on 59, and worse on 12. For *Avg* values, it is better on 65, equal on 17, and worse on 9 instances.
- ◻ ILS vs. GA2: In terms of *Best* values, ILS is better on 60 instances, equal on 21, and worse on 10. For *Avg* values, it outperforms GA2 on 69 instances, equal on 14, and worse on 8.

Finally, when compared with *Konemann_AL*, both ILS and ABC are superior in terms of both *Best* and *Avg* values across all instances.

In terms of computational time ($ATET$), ABC algorithm and ILS are overall faster than GA1 and GA2 across all 91 instances, with the exception of two cases – *g1000-1-01* ($K = 20$) and *steind5* ($K = 20$) – where GA1 and GA2 show faster execution (Tables 2.3–2.4). However, on these two instances, both ABC and ILS outperform GA1 in terms of *Avg*, and surpass GA2 in both *Best* and *Avg* solution quality.

It is important to note that GA1 and GA2 were executed on a system with a 3.40 GHz Intel Core *i*7-4770 processor and 8 GB RAM, as reported in [34], whereas ABC algorithm and ILS were run on a system with a 3.2 GHz × 4 Intel Core *i*5 processor and 4 GB RAM.

In summary, the results of ABC algorithm and ILS on 49 BX weighted instances and 42 BB weighted instances highlight its superior robustness compared to GA1 and GA2, in both solution quality (*Avg*) and computational time ($ATET$). Additionally, ABC algorithm and ILS identify new *Best* values in 17 out of 91 instances and 18 out of 91 instances as compared to GA1 and GA2, as indicated by the instances highlighted in bold in Tables 2.3-2.4. Both ABC algorithm and ILS are overall computationally faster ($ATET$) than GA2 and GA2.

Between ILS and ABC algorithm, ILS performs better on 11 instances, equal on 78, and worse on 2 instances in terms of finding the *Best* value. In terms of the *Avg* value,

Table 2.3: Results of Approx_Kon, GA1, GA2 , ABC algorithm and ILS on BX instances

Instance	|V|	K	*Konemann AL*	GA1				GA2				ABC					ILS				
			Value	Best	Avg	ATTB	ATET	Best	Avg	ATTB	ATET	Best	Avg	SD	ATTB	ATET	Best	Avg	SD	ATTB	ATET
g25-4-01	25	2	20.600000	10.333333	10.333333	0.033	0.607	10.333333	10.333333	0.006	0.676	10.33	10.33	0	0.00	0.01	10.33	10.33	0	≈ 0.00	≈ 0.00
		3	8.344828	8	**8**	0.022	0.773	8	**8**	0.011	0.614	8	**8**	0	0.00	0.01	8	8	0	≈ 0.00	≈ 0.00
		4	7.166667	**5.4**	**5.4**	0.216	1	**5.4**	**5.435135**	0.056	0.746	5.96	5.96	0	0.00	0.01	5.96	5.98	0.07	≈ 0.00	≈ 0.00
		5	5.965517	**5**	**5.114479**	0.198	1.112	5.142857	5.196571	0.04	0.747	5.14	5.14	0	0.00	0.01	5.14	5.14	0	≈ 0.00	≈ 0.00
		10	4.916667	2.6	2.651429	0.684	2.023	2.6	2.821022	0.406	1.713	**2.16**	**2.16**	0	0.00	0.02	**2.16**	2.34	0.18	≈ 0.00	0.01
		15	4.916667	2.111111	2.111111	0.42	2.341	2.111111	2.117111	0.556	2.746	**1.74**	**1.74**	0	0.00	0.01	**1.74**	**1.74**	0	≈ 0.00	0.01
		20	4.916667	1.823529	1.823529	0.136	2.289	1.823529	1.828186	0.192	3.377	**1.13**	**1.13**	0	0.00	0.02	**1.13**	**1.13**	0	≈ 0.00	≈ 0.00
g50-4-01	50	2	22.428571	16	16.7	0.285	3.683	16	**16.245238**	0.06	0.735	17	17	0	0.00	0.03	17	17	0	≈ 0.00	0.01
		3	14.428571	11.222222	11.95235	1.04	4.485	11.222222	11.911111	0.098	0.909	11.73	**11.73**	0	0.00	0.04	11.73	11.76	0.11	≈ 0.00	0.02
		4	14.428571	9.588235	10.623648	0.731	4.421	9.588235	10.73225	0.142	1.019	10.27	**10.29**	0.07	0.01	0.04	10.27	10.39	0.22	0.01	0.03
		5	14.428571	9	9.177745	1.105	4.672	9	9.618384	0.203	1.272	9	**9.05**	0.15	0.01	0.05	9	**9.02**	0.07	0.01	0.03
		10	8.708333	5	5.751302	2.267	6.623	5	6.118029	1.073	4.099	5.46	**5.46**	0	0.03	0.11	5.46	**5.49**	0.07	0.03	0.08
		15	7.166667	4.234043	4.55421	2.623	9.184	4.234043	4.531934	2.468	7.547	4.32	**4.32**	0	0.04	0.15	4.32	**4.4**	0.14	0.03	0.09
		20	5.363636	3.195652	3.655856	3.774	11.288	3.328358	3.781173	3.824	10.986	3.29	**3.32**	0.04	0.11	0.3	3.2	**3.28**	0.08	0.04	0.10
g75-4-01	75	2	27.750000	22.666667	22.666667	0.303	4.149	22.666667	23.214583	0.156	1.287	22.67	**22.67**	0	0.01	0.09	22.67	22.67	0	0.01	0.04
		3	31.750000	17.75	17.7625	0.753	4.732	17.75	17.964167	0.137	1.837	17.8	17.8	0	0.01	0.09	17.8	17.8	0	0.02	0.07
		4	19.666667	16.2	16.2025	0.143	4.397	16.2	16.4075	0.122	2.101	16.2	**16.2**	0.01	0.02	0.1	16.2	16.2	0.01	0.03	0.09
		5	18.333333	11.666667	12.420449	0.722	4.257	11.666667	13.66903	0.643	2.951	11.67	**11.67**	0	0.04	0.12	11.67	11.93	0.53	0.03	0.11
		10	11.047619	7.666667	8.489212	2.507	7.629	8.304348	9.260038	1.086	6.075	7.67	**7.94**	0.5	0.09	0.27	7.67	8.11	0.76	0.07	0.26
		15	10.740741	5.466667	6.468417	4.575	11.043	5.466667	6.425334	3.212	10.324	5.47	**5.69**	0.1	0.17	0.45	5.47	**5.62**	0.18	0.15	0.34
		20	9.000000	4.116279	4.786497	6.772	14.976	4.482143	4.969506	11.035	19.892	4.33	**4.59**	0.18	0.36	0.86	**4.24**	**4.33**	0.13	0.24	0.43
g100-4-01	100	2	33.333333	30.25	30.25	0.025	3	30.25	30.404167	0.008	1.562	30.25	30.25	0	0.00	0.17	30.25	30.25	0	0.00	0.10
		3	29.230769	24.333333	24.566666	0.792	4.879	24.555556	24.977778	0.019	2.348	24.33	**24.43**	0.24	0.04	0.18	24.33	**24.33**	0	0.04	0.16
		4	29.230769	13.333333	14.229982	0.644	4.946	14	17.560806	0.41	2.926	14	**14.04**	0.18	0.05	0.19	14	**14.08**	0.25	0.06	0.21
		5	15.500000	12.24	13.3365	0.745	5.44	13.333333	14.051923	0.458	3.508	12.24	**13.01**	0.5	0.04	0.2	12.24	**13.22**	0.33	0.04	0.25
		10	12.538462	8.9375	9.747968	3.383	10.062	9.307692	10.117313	1.573	9.162	8.94	**9.14**	0.15	0.22	0.57	8.94	**9.21**	0.22	0.32	0.70
		15	13.461538	7.375	8.114372	6.535	15.902	7.545455	8.262583	4.019	16.56	7.38	**7.61**	0.18	0.33	0.87	7.38	**7.68**	0.23	0.32	0.80
		20	8.526316	6.052632	7.159488	9.689	21.323	5.931818	7.09543	8.284	25.843	6.26	**6.47**	0.14	0.83	1.84	6.17	**6.29**	0.23	0.48	1.07
g200-4-01	200	2	78.000000	40.142857	40.142857	0.245	9.493	40.142857	41.125714	0.333	6.945	40.14	40.14	0	0.04	1.04	40.14	40.14	0	0.04	0.41
		3	31.800000	31.8	31.8	0.49	11.786	31.8	32.365714	0.283	9.298	31.8	31.8	0	0.05	1.04	31.8	31.8	0	0.08	0.94
		4	31.800000	27.111111	28.811111	1.039	13.348	29	29	1.393	11.477	27.11	**27.96**	0.94	0.27	1.13	27.11	**27.3**	0.57	0.37	1.70
		5	31.923077	22.454545	23.662093	2.353	15.649	23	26.813889	4.445	15.804	22.6	**22.62**	0.09	0.31	1.14	22.6	**22.6**	0	0.44	2.01
		10	27.142857	16.6	17.532976	5.991	25.666	17	18.521628	12.229	32.072	**15.62**	**16.22**	0.42	1.78	3.8	**15.62**	**16.19**	0.41	2.66	5.70
		15	20.000000	11	12.805148	17.528	40.44	11.181818	14.953014	11.752	40.52	**10.62**	**11.01**	0.17	2.18	5.06	10.71	11.02	0.22	3.57	7.21
		20	13.461538	10.393939	11.074202	19.736	48.763	10.685714	11.22531	18.098	56.671	**10.09**	**10.41**	0.18	5.45	11.63	**9.4**	**10.17**	0.43	4.38	9.37
g400-4-01	400	2	441.000000	93	**93**	1.648	36.537	93	157.1	5.895	31.497	93	**93**	0	0.34	7.3	93	93	0	0.77	2.47
		3	91.000000	71.5	**71.5**	5.497	47.988	72	82.908333	8.562	41.13	71.5	**71.5**	0	0.93	7.17	71.5	71.5	0	0.95	5.86
		4	91.000000	55	55.375	8.245	52.675	55.666667	66.233333	7.844	43.916	55	**55**	0	1.77	7.24	55	55.08	0.33	3.11	10.85
		5	91.000000	51.4	54.64	10.132	60.328	55	65.275	6.457	46.797	51.4	**53.56**	1.76	1.64	7.56	51.4	54.28	1.44	2.27	17.41
		10	54.000000	29	32.121905	32.156	95.382	31.5	38.248571	22.096	70.535	**25.86**	**29.48**	1.11	12.03	23.85	25.86	**29.26**	1.02	28.36	48.66
		15	47.444444	21	24.448889	41.624	119.684	24.6	29.68627	31.325	91.147	21.1	**22.34**	0.47	20.40	38.73	20.74	**22.02**	0.58	32.40	62.35
		20	25.400000	17.769231	20.512649	95.592	193.912	19.4	25.071429	30.828	105.145	17.8	**19.05**	0.65	41.22	85.04	15.56	**17.17**	0.7	44.54	76.47
g1000-4-01	1000	2	196.000000	138	138.2	38.195	467.519	143	159.9	29.203	224.296	138	**138**	0	13.94	102.3	138	138.55	1.32	5.02	16.09
		3	196.000000	114	114.4	33.021	483.743	122	142.05	40.486	248.211	114	**114**	0	10.26	100.7	114	114.4	1.74	10.70	35.92
		4	143.000000	90	94.05	121.116	558.249	107.666667	121.35	84.462	284.886	90	**90.45**	1.96	39.83	110.94	90	93.3	5.72	24.41	63.80
		5	143.000000	78	88.05	160.485	594.203	90	110.85	94.691	521.564	**76.33**	**80.48**	4.93	63.52	142.05	76.33	83.54	5.53	41.00	100.13
		10	118.750000	55	63.008333	371.548	891.103	70.5	79.366667	263.645	758.729	**54.33**	**57.01**	2.2	182.29	356.4	**53**	57.01	2.44	214.77	378.38
		15	102.000000	45.666667	55.4	330.472	908.486	57	65.8	241.235	814.247	**43.67**	**45.27**	1.81	358.84	690.12	**43**	**45.03**	2.77	541.82	837.73
		20	70.833333	38.5	46.197619	429.589	988.091	53	64.7	276.545	905.573	38.5	**40.5**	1.33	620.23	1395.73	**35**	**39.39**	4.15	791.31	1339.16

Table 2.4: Results of *Konemann_AL*, GA1, GA2, ABC algorithm and ILS on BB weighted instances

Instance	|V|	K	*Konemann_AL*	GA1				GA2				ABC					ILS				
			Value	Best	Avg	ATTB	ATET	Best	Avg	ATTB	ATET	Best	Avg	SD	ATTB	ATET	Best	Avg	SD	ATTB	ATET
15 x 15	255	2	855.000000	403	**403**	0.226	12.207	423	423	0.019	6.137	403	**403**	0	0.05	1.22	403	403	0	0.05	0.50
		3	247.000000	157	159.4	1.636	15.148	167	176.05	0.217	9.865	157	**157**	0	0.15	1.23	157	**157**	0	0.23	1.13
		4	209.000000	134	142.4	2.071	15.952	148	152.35	0.517	11.676	134	**138.88**	6.64	0.39	1.31	134	**136.1**	5	0.42	2.03
		5	209.000000	98	100.508333	3.722	18.017	98	116.593333	7.432	21.248	98	**98**	0	0.38	1.29	98	**98**	0	0.41	2.72
		10	81.400000	50.555556	53.266667	7.223	28.85	51	57.5	10.516	29.562	50.56	**50.67**	0.19	0.74	2.31	50.56	**50.67**	0.19	1.17	7.40
		15	57.375000	25	31.906667	14.364	41.638	26.25	36.463373	11.321	36.433	**24.71**	**25.16**	0.56	2.85	5.18	**24.71**	26.15	3.49	3.73	10.02
		20	62.714286	20.666667	23.19	20.656	53.142	21	23.460833	22.303	52.267	**18.33**	**19.71**	0.75	6.76	12.47	**18.33**	**18.53**	0.62	4.85	13.63
33 x 33	1089	2	1863.000000	1087	**1088**	105.541	772.058	1087	1134.3	32.731	591.818	1087	1091.2	7.53	31.44	117.11	1087	1088.4	4.25	7.17	17.74
		3	1863.000000	707	**707**	159.031	910.835	707	743	83.901	673.265	707	**707**	0	23.2	116.78	707	707	0	13.60	38.97
		4	931.000000	609	613.7	363.148	1060.107	643	688.7	208.041	808.344	609	**611**	4	66.57	131.73	609	616.1	14.72	34.13	71.54
		5	795.000000	601	607	195.902	1012.91	595	628.4	148.951	821.428	595	**600.6**	4.13	44.34	129.09	601	604.8	4.28	28.52	107.37
		10	581.000000	319	368.35	650.386	1354.273	343	383.7	222.947	968.63	319	**327.1**	9.01	265.81	415.42	319	330.6	12.57	196.20	425.22
		15	423.000000	195	258.416667	902.792	1548.094	227	284.316667	383.558	1130.592	195	**211.9**	9.65	419.97	645.36	195	**208.2**	13.48	479.05	892.73
		20	281.000000	185	206.15	741.422	1519.992	185	215.15	348.991	1149.519	**148**	**168.1**	6.61	890.89	1470.59	139	**163.5**	21.42	900.58	1516.48
45 x 5	225	2	963.000000	483	**483**	0.293	12.429	483	**483**	0.085	10.03	483	**483**	0	0.04	1.23	483	483	0	0.05	0.50
		3	337.000000	208.333333	**208.333333**	0.638	15.07	208.333333	246.6	2.285	14.231	208.33	**208.33**	0	0.12	1.22	208.33	208.33	0	0.13	1.12
		4	337.000000	183	**183**	1.549	16.77	183	198.533333	3.478	16.429	183	**183**	0	0.15	1.23	183	183	0	0.19	2.00
		5	337.000000	146.333333	150.9	3.222	18.221	146.333333	175.033333	1.976	17.122	146.33	**146.83**	2.18	0.62	1.49	146.33	**146.33**	0	0.95	2.86
		10	85.666667	59	64.926667	4.193	23.06	61	73.94	9.116	25.9	59	**59.1**	0.44	0.83	2.39	59	59.8	2.4	0.70	7.29
		15	49.000000	32	37.15	8.848	32.795	34	40.733333	14.371	40.808	32	**33.03**	1.25	2.02	4.33	32	33.3	1.1	2.31	9.88
		20	38.000000	19	25.608333	19.13	44.755	395	473.3	2260.595	6677.6	**17.67**	**18.54**	0.63	7.31	12.96	**17.67**	19.27	3.95	5.83	13.73
100 x 10	1000	2	3037.000000	1689	**1689**	35.178	484.535	1689	1733	22.128	271.1	1689	**1689**	0	7.71	96.51	1689	1692.6	7.36	5.98	14.61
		3	1497.000000	1041	1041.1	167.014	623.227	1121	1146.3	30.088	351.567	1041	**1041**	0	22.38	90.32	1041	1041.8	1.17	17.89	33.69
		4	1399.000000	867	**869.1**	163.167	623.724	937	1041	59.017	366.924	867	**869.1**	5	39.58	96	867	869.4	5.04	28.65	59.15
		5	1293.000000	703	718.7	127.257	626.312	743	831.4	41.486	388.885	703	**703.6**	2.62	41.57	103.33	703	706.4	13.92	31.74	92.00
		10	633.000000	395	405.1	194.575	760.973	397	487.1	88.935	492.907	**375**	**391.9**	8.8	111.23	228.9	**375**	**389.9**	8.31	140.97	351.61
		15	527.000000	265	297.3	333.968	886.385	326	359	169.162	604.22	**259**	**269.35**	10.75	331.72	507.5	**259**	269.8	14.81	269.24	718.41
		20	347.000000	189	228.123333	532.296	1062.399	223	261.4	173.97	647.442	**183**	**194.6**	6.91	667.07	1120.4	**161**	**192.2**	19.35	593.28	1182.56
steinc5	500	2	463.000000	257	257.7	10.194	75.483	261	278.6	5.967	54.861	257	**257**	0	0.52	12.4	257	257	0	0.50	3.78
		3	537.000000	203	204.6	13.131	82.895	221	251.1	14.068	72.982	203	**203**	0	2.26	12.47	203	203.35	0.91	4.07	8.93
		4	357.000000	179	182.85	8.911	83.491	186	218.85	5.656	68.237	179	**179.35**	1.53	4.82	12.61	179	180.05	2.5	3.92	15.19
		5	275.000000	169	170.55	24.268	103.021	176	200.45	24.627	93.12	169	**169**	0	4.71	13.18	169	169	0	4.42	23.74
		10	135.000000	88	96.75	38.433	138.411	99	129.275	48.709	136.838	88	**89.35**	3.21	13.42	29.58	88	94.3	4.12	14.37	85.29
		15	97.000000	52.428571	62.527381	88.057	199.472	61.5	97.975	56.565	161.846	52.43	**54**	2.54	43.43	66.93	52.43	**53.38**	2.19	80.41	131.10
		20	100.000000	47.666667	54.978571	88.144	231.482	49.571429	68.400794	71.507	200.101	47.67	**48.53**	0.75	67.43	127.67	47.67	54.92	22.34	48.70	157.06
steind5	1000	2	815.000000	473	**473**	27.734	485.418	509	538.8	50.692	395.576	473	**473**	0	4.7	90.23	473	473	0	2.91	15.34
		3	795.000000	457	457.1	50.096	551.608	459	477.5	33.335	477.755	457	**457**	0	16.53	90.75	457	457	0	10.53	34.58
		4	737.000000	367	378.8	144.643	652.206	435	449.4	72.868	504.82	367	**374.2**	11	28.1	95.95	367	376.6	11.76	18.66	61.92
		5	541.000000	349	351.6	75.588	606.527	367	413.8	82.437	530.674	349	**349.2**	0.87	22.47	92.58	349	349.6	1.43	23.03	94.51
		10	425.000000	197	237.4	341.623	901.829	271	316.9	205.511	688.88	197	**213**	12.76	161.52	280.12	197	224.4	26.22	94.56	341.43
		15	303.000000	139	162.7	630.996	1072.29	201	247.9	217.661	747.034	139	**146.85**	5.49	271.93	450.91	139	**140.75**	4.09	371.40	730.54
		20	201.000000	117	137.85	551.433	1123.258	145	204.6	319.596	854.022	117	**119.11**	3.27	705.07	1166.8	117	122.5	9.6	531.12	1246.57

ILS is better on 26 instances, equal on 27, and worse on 38 instances compared to ABC algorithm.

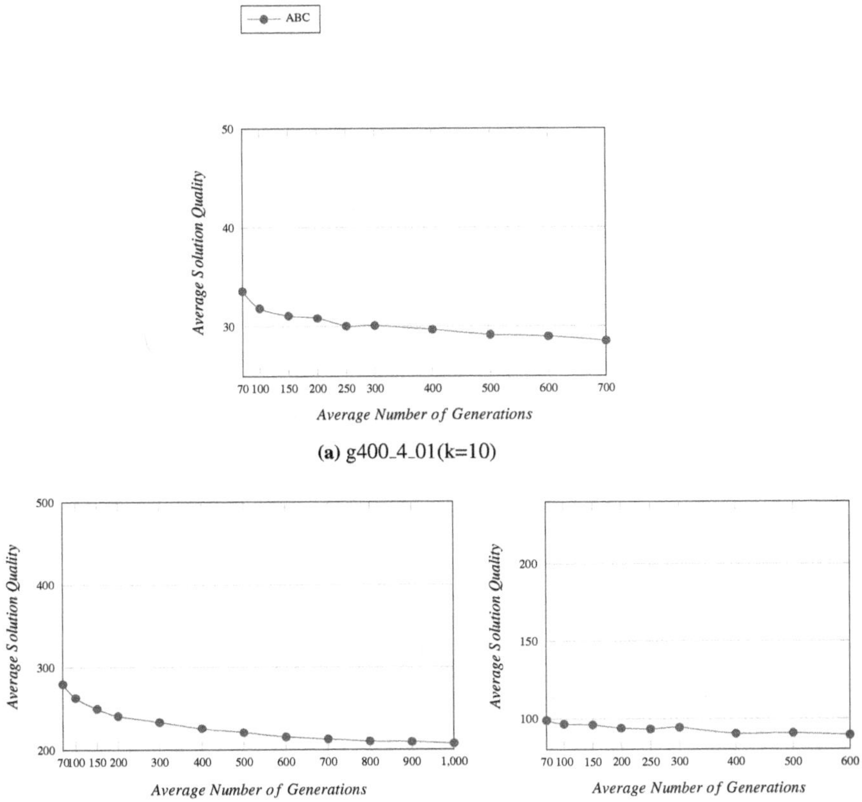

(a) g400_4_01(k=10)

(b) bb33x33_1(k=15)

(c) steinc5(k=10)

Figure 2.1: Evolution of average solution quality over average number of generations for ABC Algorithm

2.6.4 Convergence behavior of ABC algorithm

This subsection discusses the convergence behavior of ABC algorithm. To carry out this experimentation, three instances < g400_4_01 (k=10), bb33x33_1 (k=15), steinc5 (k=10)> are considered. Figures 2.1 (a-c) illustrate the evolution of solution quality (*Avg*) against the successive average number of generations. The X-axis represents the "Average Number of Generations" over 20 runs, while the Y-axis represents the "Average Solution Quality" over 20 runs. It is noteworthy that we opted for successive average number of generations instead of average total execution time (ATET) or the termination criterion of ABC algorithm, as the total number of generations excuted within a given termination criterion of ABC algorithm can be easily computed. The curves in Figures 2.1 (a-c) depict a continuous decrease in the fitness of the solution obtained by ABC algorithm with the increase in generations.

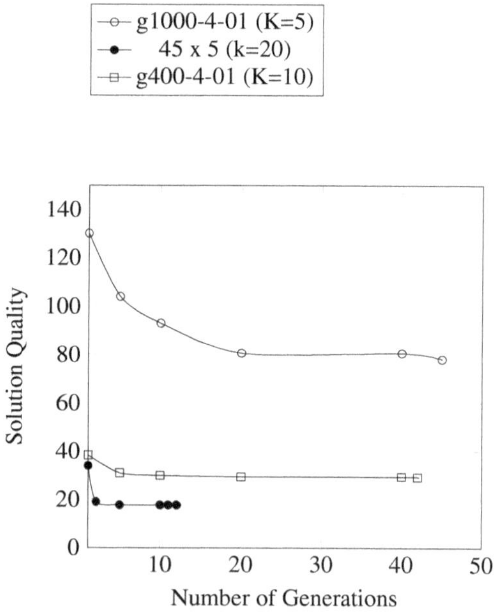

Figure 2.2: Evolution of *Solution Quality* against *Number of Generations* for ILS Algorithm

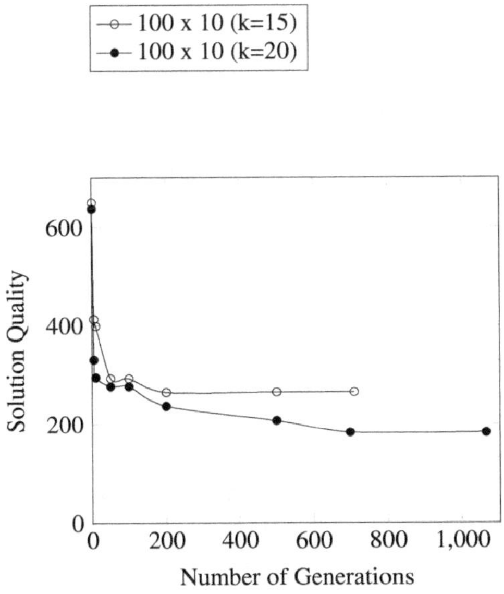

Figure 2.3: Evolution of *Solution Quality* against *Number of Generations* for ILS Algorithm

2.6.5 Convergence behavior of ILS

This subsection discusses the convergence behavior of ILS. For this, we have considerd 5 different instances (45 x 5 (k=20), g400-4-01 (K=10), g1000-4-01 (K=5), 100 x 10 (k=15), 100 x 10 (k=20)) whose sizes vary from 225 vertices to 1000 vertices. Figure 2.2 which is for the first three instances (45 x 5 (k=20), g400-4-01 (K=10), g1000-4-01 (K=5) and Figure 2.3 which is for the last two instances (100 x 10 (k=15), 100 x 10 (k=20) illustrate the evolution of solution quality on single run against successive number of generations, where X-axis indicates the "Number of Generations", while Y-axis indicates the "Solution Quality". We have considered the same termination criterion of ILS explained in Section 5. Curves in Figures 2.2-2.3 show that the fitness of solution obtained by ILS decreases continuously with the increase of generations. One can observe from the Figures that ILS, based on the termination criterion, converges

Table 2.5: Results of ABC_Nbr_cv, ABC_Nbr_swp, and ABC algorithm on some benchmark instances

Instance	$\|V\|$	K	ABC_Nbr_cv		ABC_Nbr_swp		ABC	
			Avg	ATET	Avg	ATET	Avg	ATET
g25-4-01	25	10	2.21	0.01	2.22	0.02	**2.16**	0.02
g50-4-01	50	10	5.67	0.07	5.47	0.13	**5.46**	0.08
		15	4.37	0.09	4.33	0.15	**4.32**	0.1
g75-4-01	75	4	16.39	0.04	**16.2**	0.18	**16.2**	0.08
		5	13.05	0.06	11.73	0.19	**11.67**	0.1
		20	4.62	0.36	4.79	1.25	**4.59**	0.62
g200-4-01	200	3	31.97	0.29	**31.8**	2.5	**31.8**	0.89
		10	18.33	0.98	16.27	7.43	**16.22**	2.99
g400-4-01	400	3	73.03	1.65	**71.5**	19.41	**71.5**	6.47
		4	62.56	1.77	**55**	19.61	**55**	6.48
g1000-4-01	1000	2	139.3	17.33	**138**	292.42	**138**	92.79
		3	116.43	12.31	**114**	293.46	**114**	90.88
15x15	255	2	411.6	0.33	**403**	3.52	**403**	1.22
		3	167.1	0.39	**157**	3.51	**157**	1.23
		5	108.26	0.45	**98**	3.56	**98**	1.29
33x33	1089	3	718.6	14.11	**707**	379.51	**707**	116.78
		10	358.95	36.94	334.55	484.45	**327.1**	415.42
		15	244.83	66.49	224.35	830.83	**211.9**	645.36
		20	184.02	95.05	168.27	1822.32	**168.1**	1470.59
45x5	225	2	483	0.33	**483**	3.63	**483**	1.23
		20	20.24	2.64	19.43	36.22	**18.54**	12.96
100x10	1000	2	1689	23.32	**1689**	300.97	**1689**	96.51
		3	1058	13.65	**1041**	302	**1041**	90.32
		5	757.3	13.31	703.8	117.55	**703.6**	103.33
		10	405.4	30.55	398.6	268.4	**391.9**	228.9
		15	290.42	48.83	273.05	830.52	**269.35**	507.5
steinc5	500	2	257.4	3.52	**257**	38.35	**257**	12.4
		3	211.7	2.93	**203**	38.39	**203**	12.47
		5	177.6	2.85	**169**	38.87	**169**	13.18
		10	99.8	6.85	**89.35**	87.89	**89.35**	29.58
steind5	1000	2	480.2	11.11	**473**	300.87	**473**	90.23
		3	458.7	9.81	**457**	303.1	**457**	90.75

towards better and better solution quality even the size of instance increases. One can infer from this convergence behavior of ILS that ILS is scalable even for large size of instances.

2.6.6 Impact of neighborhood operators (*Nbr_cv* and *Nbr_swp*)

Before considering the use of both neighborhood operators (*Nbr_cv* and *Nbr_swp*) in a mutually exclusive way in ABC algorithm, a preliminary experiment was conducted. Two variants, namely ABC_Nbr_cv and ABC_Nbr_swp were created by disabling *Nbr_swp* and *Nbr_cv*, respectively, in ABC algorithm. These variants were executed on some benchmark instances using a setup similar to ABC algorithm. The results, including the average solution quality (Avg) and average total execution time ($ATET$) of ABC_Nbr_cv, ABC_Nbr_swp, and ABC algorithm are reported in Table 2.5. The findings indicate that ABC algorithm, utilizing both neighborhood operators (*Nbr_cv* and *Nbr_swp*) in a mutually exclusive way, achieves better average solution quality (highlighted in bold) on most of the considered instances while requiring less computational time. It is noteworthy that the *Nbr_swp* operator is computationally more expensive compared to *Nbr_cv*. The use of both operators (*Nbr_cv* and *Nbr_swp*) in a mutually exclusive way in ABC algorithm demonstrates its effectiveness in obtaining high-quality solutions within a reasonable computational time.

2.6.7 Impact on solution quality using different values of *nrcv* in perturbation procedure

This section investigates on considering different values of *nrcv* in the perturbation procedure and examines how quality of solutions get impacted. For this, we have considered *nrcv*=2 and *nrcv*=3 in addition to *nrcv*=1. To carry out this investigation, we have considered a set of instances reported in Table 2.6. Table 2.6 also reports the results of ILS on these instances when *nrcv*=1, *nrcv*=2 and *nrcv*=3. Best results are highlighted in bold. It is clear from the results of Table 2.6 that *nrcv*=1 dominates over *nrcv*=2 and *nrcv*=3. Hence, *nrcv*=1 is the most suitable for the ILS on considered instances.

Table 2.6: Impact on solution quality using different values of *nrcv* in perturbation procedure

Instance	K	*nrcv*=1		*nrcv*=2		*nrcv*=3	
		Best	Avg	Best	Avg	Best	Avg
33 x 33	5	**601.000000**	**604.799988**	**601.000000**	605.799988	**601.000000**	608.599976
33 x 33	10	**319.000000**	**330.600006**	**319.000000**	330.750000	**319.000000**	348.000000
33 x 33	15	**195.000000**	**208.199997**	195.000000	219.350006	267.000000	286.083313
33 x 33	20	**139.000000**	**163.499985**	141.000000	166.550003	189.000000	214.000000
100 x 10	5	**703.000000**	**706.400024**	703.000000	709.799988	705.000000	740.200012
100 x 10	10	**375.000000**	**389.899994**	381.000000	394.600006	391.000000	399.299988
100 x 10	15	**259.000000**	**269.799988**	259.000000	273.100006	267.000000	286.083313
100 x 10	20	**161.000000**	**192.199997**	183.000000	208.750000	189.000000	214.000000
g400-4-01	10	**25.857143**	**29.263748**	28.142857	30.324606	28.142857	31.275951
g400-4-01	15	**20.736841**	**22.021326**	21.285715	22.504704	22.714285	23.535458
g400-4-01	20	**15.560000**	**17.169756**	16.714285	18.055241	17.350000	19.962019
g1000-4-01	5	76.333336	**83.541672**	78.000000	85.416664	76.333336	87.716667
g1000-4-01	10	**53.000000**	**57.008339**	53.666668	59.258888	59.000000	62.027382
g1000-4-01	15	**43.000000**	**45.033337**	43.666668	47.015003	45.500000	49.004166
g1000-4-01	20	**35.000000**	**39.391670**	36.000000	39.424622	38.500000	41.518333

2.6.8 Statistical Analysis

Table 2.7: Statistical test on the results of *Best* values

Approches	p-value	significant
ABC vs GA1	0.02444	yes
ABC vs GA2	<0.00001	yes

Table 2.8: Statistical test on the results of *Avg* values

Approches	p-value	significant
ABC vs GA1	0.00001	yes
ABC vs GA2	<0.00001	yes

Table 2.9: Statistical test on the results of *Best* values

Approaches	p-value	significant
ILS vs GA1	0.00424	yes
ILS vs GA2	<0.00001	yes

Table 2.10: Statistical test on the results of *Avg* values

Approaches	p-value	significant
ILS vs GA1	<0.00001	yes
ILS vs GA2	<0.00001	yes

Table 2.11: Statistical test on the results of *Best* values

Approches	p-value	significant
ABC vs ILS	0.0251	yes

For the statistical analysis, a two-tailed nonparametric Wilcoxon's signed-rank test was conducted on the best (*Best*) and average solution quality (*Avg*) values across 91 benchmark instances (49 BX instances and 42 BB instances, as reported in Tables 2.7-2.10) by using Wilcoxon's Signed-Rank test calculator [69]. The significance criterion

Table 2.12: Statistical test on the results of *Avg* values

Approches	p-value	significant
ABC vs ILS	0.09692	no

was set at 0.05. The results of Wilcoxon's signed rank test, reported in Tables 2.7-2.8 and Tables 2.9-2.10, show that ABC algorithm and ILS find a significant difference with GA1 and GA2. On the best value (*Best*), there is a significant difference between ABC algorithm and ILS (see Table 2.11), while there is no significant difference between ABC algorithm and ILS (see Table 2.12).

Chapter 3

Tree t-spanner problem

3.1 Introduction

Let $G\ (V, E, w)$ be an edge-weighted, undirected and connected graph, where V represents the vertex set, E represents the edge set, and a positive weight w is associated with each edge. A spanning tree (say Tr) of G is referred to as a tree t-spanner if the ratio of the shortest distance between any pair of vertices (say p and q) in Tr and G is at most t among all pair of vertices in V. The value of t is known as the stretch factor of Tr, and the ratio (say $d_{Tr}(p,q)/d_G(p,q)$) is called the stretch, where $d_{Tr}(p,q)$ and $d_G(p,q)$ respectively represent a shortest distance between p and q vertices in Tr and in G. The tree t-spanner problem (Tr_t-SP) for a given G aims to identify a spanning tree having minimum t over all spanning trees of G.

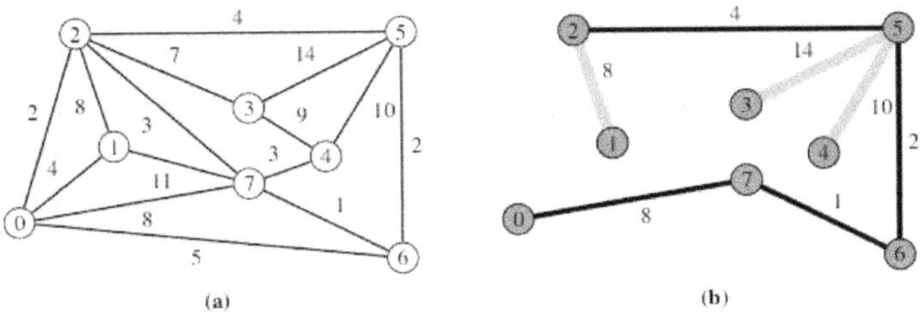

Figure 3.1: A graph (G) and a spanning tree (Tr) of G

(p,q)	$d_{Tr}(p,q)$	$d_G(p,q)$	$\frac{d_{Tr}(p,q)}{d_G(p,q)}$
(0,1)	23.00	4.00	5.75
(0,2)	**15.00**	**2.00**	**7.50**
(0,3)	25.00	9.00	2.78
(0,4)	21.00	8.00	2.62
(0,5)	11.00	6.00	1.83
(0,6)	9.00	5.00	1.80
(0,7)	8.00	5.00	1.60
(1,2)	8.00	6.00	1.33
(1,3)	26.00	13.00	2.00
(1,4)	22.00	12.00	1.83
(1,5)	12.00	10.00	1.20
(1,6)	14.00	9.00	1.56
(1,7)	15.00	9.00	1.67
(2,3)	18.00	7.00	2.57
(2,4)	14.00	6.00	2.33
(2,5)	4.00	4.00	1.00
(2,6)	6.00	4.00	1.50
(2,7)	7.00	3.00	2.33
(3,4)	24.00	9.00	2.67
(3,5)	14.00	11.00	1.27
(3,6)	16.00	11.00	1.45
(3,7)	17.00	10.00	1.70
(4,5)	10.00	6.00	1.67
(4,6)	12.00	4.00	3.00
(4,7)	13.00	3.00	4.33
(5,6)	2.00	2.00	1.00
(5,7)	3.00	3.00	1.00
(6,7)	1.00	1.00	1.00

Table 3.1: The stretch of each pair of vertices (p,q) for Tr depicted in Figure. 3.1(b)

To understand this problem, Figure 3.1 (a-b) illustrate a graph (G) as an input and a spanning tree (Tr) of G. Table 3.1 reports the stretch of each pair of vertices (p,q) for Tr shown in Figure 3.1 (b). From the given Table 3.1, one observes that among all pair of vertices, the pair $(0,2)$, which is highlighted in bold, has the maximum stretch, and hence its stretch factor (t) is 7.5 (highlighted in bold).

Peleg and Ullman [70] were the first who coined the term *"t-spanner"* and defined it as a sub-graph of a given graph if the ratio of the shortest distance between any pair of vertices in the subgraph and the given graph is at most t among all its pair of vertices. They explained that such spanners can be used for building network synchronizers that allow any synchronous algorithm to execute on any asynchronous network [70]. The property of spanner, based on approximating such ratio, also allows spanners to be useful in several fields such as distributed systems, communication networks, biology involving the rebuilding phylogenetic trees from matrices whose entries represent genetic distances among living species, robotic systems, network design [53, 70, 71, 72, 73, 74].

A spanning subgraph T_r in a connected graph can be called a tree t-spanner if T_r holds both properties of t-spanner and a spanning tree. Such a spanning tree is also called a tree-spanner, and a tree-spanner is called a tree t-spanner with $t \geq 1$. A tree spanner with low stretch has its practical advantages, as it simplifies routing [53].

3.1.1 Literature Review

In the literature, the pioneering work of Cai and Corneil [75] made important contributions towards theoretical and algorithmic problems associated with tree spanners of edge-weighted and unweighted connected graphs. They concluded that a tree 1-spanner (t=1), if it exists, is a minimum spanning tree and it can be achieved in polynomial time; otherwise, the tree t-spanner problem is $\mathcal{NP}$-hard for any fixed number $t > 1$ for connected, undirected and weighted graph. They also concluded that the tree t-spanner problem is $\mathcal{NP}$-hard for any fixed number $t \geq 4$ on unweighted connected graphs. Note that for connected, undirected and unweighted graphs, researchers also referred to the tree t-spanner problem in the literature as the minimum max-stretch spanning tree problem [76]. Authors [76] proposed an $\mathcal{O}(\log n)$ -approximation algorithm for unweighted graphs. Moreover, several works on developing approximation algorithms for some specialized kinds of unweighted graphs [77, 78, 79, 80, 81, 82] have been accomplished. Recently, some theoretical advancements on this problem have been made, particularly when focusing on specific graph classes [83, 84, 85]. Kardam *et al.* [86] presented a metaheuristic technique based on general variable neighborhood search for this problem on connected, undirected and unweighted graph (unit edge-weight). Recently, Couto *et al.* [87] worked on two exact brute-force algorithms (sequential and parallel) and also proposed two greedy heuristics for this problem on unweighted graphs.

The tree t-spanner problem (Tr_t-SP) associated with connected, undirected, and edge-weighted graphs is the topic of this study. Due to its computational intractability, developing metaheuristic techniques for such problems to determine high quality solutions within an acceptable computational time has attracted much attention in the literature. In this direction, several metaheuristic techniques for the Tr_t-SP have been proposed. For the Tr_t-SP, Moharam and Morsy [88] presented a genetic algorithm; Israni and Sundar [89] presented ant colony optimization approach; Sundar [90] presented a steady-state genetic algorithm; and Singh and Sundar [36] presented an artificial bee colony (ABC) algorithm. Among all the proposed metaheuristic techniques, the performance of ABC algorithm is superior in terms of computational results on available benchmark instances. Moreover, Eduardo and Markus [91] recently presented a branch-and-cut exact approach based on mixed-integer programming formulations, testing it on small-sized instances with $|V| = \{10, 15, 20\}$, where the procedure for generating instances is similar to that of [36]. They found the Tr_t-SP to be quite challenging even for their proposed exact approach.

3.1.2 Our Contribution

The studies mentioned above have significantly contributed to advancing the state-of-the-art metaheuristic techniques for solving the Tr_t-SP and achieving better solution quality on available benchmark instances. It is notable that all existing metaheuristic techniques for solving Tr_t-SP are population-based metaheuristic techniques. To the best of our knowledge, no one has worked on single solution-based iterated local search (ILS) AND variable neighborhood search (VNS) approaches for the Tr_t-SP. Nevertheless, both ILS and VNS have demonstrated success in finding high-quality solutions for several $\mathcal{NP}$-hard problems.

Motivated by this, this chapter proposes two approaches: iterated local search (ILS) and Basic variable neighborhood search (VNS), for the Tr_t-SP. Within the ILS framework, we introduce a perturbation strategy to modify the starting solution, followed by a devised local search strategy to find better solutions around the perturbed state. Furthermore, a two-step strategies are devised in the acceptance criterion of ILS for the Tr_t-SP. For VNS, we propose a standard Variable Neighborhood Search for the Tr_t-SP. In this VNS framework, we introduce three neighborhood operators, i.e. N_k for $k = 1, \ldots, k_{max}$ (where $k_{max} = 3$) defining both the neighborhood structures and their order of examination. Additionally, a local search operator is applied to refine the quality of solutions generated through each neighborhood structure. A restart criterion is also implemented to increase the chances of discovering new high-quality solutions. Experimental results indicate the effectiveness of both our proposed approaches (ILS and VNS) against the existing state-of-the-art ABC algorithm on available benchmark instances.

The remaining chapter is organized as follows. The next section discusses the proposed

ILS for the Tr_t-SP; Section 3.3 discusses the proposed VNS for the Tr_t-SP; and Section 3.4 discusses the computational results.

3.2 Iterated local search for the Tr_t-SP

This section details an iterated local search (ILS) approach for the Tr_t-SP.

The proposed ILS framework for the Tr_t-SP, outlined in Algorithm 9, is composed of four key components: (a) initial solution generation, (b) a perturbation strategy, (c) a local search improvement strategy, and (d) an acceptance criterion.

Algorithm 9: The pseudo-code of ILS framework for the Tr_t-SP

Generate an initial solution Tr;
$Tr \leftarrow LS\_ImprovementStrategy(Tr)$;
while A stopping condition **do**
    $Tr' \leftarrow Perturbation(Tr)$;
    $Tr^* \leftarrow LS\_ImprovementStrategy(Tr')$;
    $Tr \leftarrow Acceptance\_Criterion(Tr, Tr^*)$;

Prior to initiating ILS for the Tr_t-SP, we pre-compute shortest path between all pairs of vertices for the specified graph instance. Subsequently, the following subsections explain the key components of the proposed ILS framework for the Tr_t-SP.

3.2.1 Solution encoding

To represent the solution for a given graph $G(V, E)$ of the Tr_t-SP as a spanning tree, we have chosen to use an edge-set encoding, as proposed by Raidl and Julstrom [92]. This encoding method comprises $|V|$-1 edges, which represent a spanning tree for the graph $G(V, E)$. This edge-set encoding can be implemented with a two-dimensional array, where the first dimension's size is set to $|V|$-1 to maintain the number of edges in a spanning tree, and the second dimension's size is 2 to accommodate the pairs of vertices defining each edge. A spanning tree (solution) Tr represented in Figure 3.1 (b) for a given input graph G represented in Figure 3.1 (a) is composed of 8-1 = 7 edges, where 8 is the total number of vertices in G. Table 3.1 displays a two dimensional array (denoted as $T[|V|-1][2]$) that implements an edge-set encoding for the spanning tree Tr of a given graph G with 8 vertices (see Figure 3.1 (a-b)).

This representation has the benefits of applying problem-specific knowledge used in the perturbation strategy as well as in the local search improvement strategy for the *Tr_t*-SP.

T[|V|-1][2]

Index number of 1st dimension	1st Index of 2nd dimension	2nd Index of 2nd dimension
0	0	7
1	2	5
2	1	2
3	6	7
4	5	6
5	3	5
6	4	5

Figure 3.2: Implementation of edge-set encoding of Tr of G through a two-dimensional array

3.2.2 Initial solution

To generate an initial solution for the *Tr_t*-SP, our procedure follows Prim's algorithm [93], except that in each iteration, instead of selecting an edge of minimum cost in G, the procedure randomly selects an edge from G that joins a vertex in the partial tree (say Tr) to a vertex not yet included in Tr, where the partial tree (say Tr) signifies that it does not yet span all vertices of the graph G. Subsequently, a local search improvement strategy is applied to this initial solution (spanning tree Tr) in order to intensify the search around it, aiming to obtain a solution of high quality. The detail of this local search improvement strategy are outlined in Section 3.2.5. This way of combining the initial solution with local search yields a better starting solution for ILS to search.

3.2.3 Computation of the fitness of a solution

The fitness (say $f(Tr)$) of a solution (referred to as the tree t-spanner, denoted as Tr) is determined by the stretch factor t, which is based on computing a stretch with maximum value (say max) over each pair of vertices (p, q) in the graph G, defined as $max\{d_{Tr}(p, q)/d_G(p, q), \forall(p, q) \in G\}$.

The procedure of computing $d_{Tr}(p, q)$ for each pair of vertices (p, q) is as follows. First, we preprocess the lowest common ancestor for each pair of vertices (p, q) in Tr with the root vertex (say *root*) by utilizing the least common ancestor algorithm [94]. Here, the lowest common ancestor for any pair of vertices, denoted as p and q, in d_{Tr} is identified as the common ancestor of vertices p and q that is farthest from the root vertex (*root*), referred to as $LCA(p, q)$. This way helps in retrieving a lowest common ancestor query in constant time for any pair of vertices. $d_{Tr}(p, q)$ between any two vertices (p and q) in Tr can be computed as follows:

$$d_{Tr}(p, q) = d_{Tr}(root, p) + d_{Tr}(root, q) - 2 \times (d_{Tr}(root, LCA(p, q)))$$

For Figure 3.1(b), the fitness of the solution is 7.5 based on the maximum stretch for (0,2) pair of vertices reported in Table 3.1.

It is to be noted that the edges (<0-7>, <7-6>, <6-5>, <5-2>) that are part of $d_{Tr}(0, 2)$ (distance between the vertex 0 and the vertex 2 in Tr) are referred to as *potential edges* of Tr highlighted in bold edges between 0 and 2 vertices in Figure 3.1(b).

3.2.4 Perturbation strategy

The role of perturbation strategy is to empower ILS to continue its search in the search space. Our proposed perturbation strategy (say *Perturbation*(Tr) procedure) shifts the search away from its current position of the solution (Tr) through performing two times (*para_pert*) (one-after-another) edge-deletion and edge-addition on the current solution, where *para_pert* is a parameter. In this procedure, a potential edge of Tr (refer to Section 3.2.3 for the potential edge) is randomly selected and deleted, causing Tr

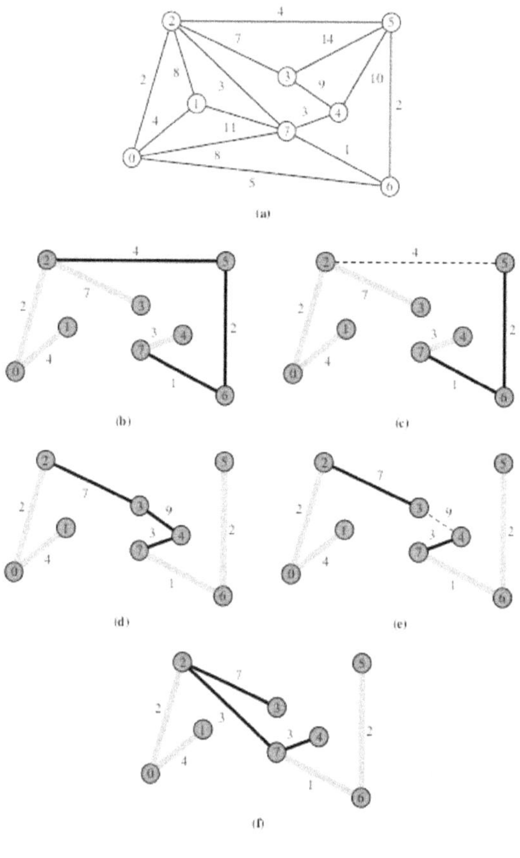

Figure 3.3: Perturbation strategy

to become an infeasible solution, as this partitioning divides Tr into two components. Here, a potential edge of a given solution is an edge that participates in its fitness. To make Tr feasible, an edge that is picked randomly from all candidate edges except the deleted edge in graph edges (E) is added to Tr, making Tr feasible, where a candidate edge means the edge that can join these two components of Tr. The procedure is repeated once more, and the resultant spanning tree becomes Tr' after applying two times (one-after-another) edge-deletion and edge-addition on Tr. Note that after performing the first round of an edge-deletion and edge-addition, the fitness of feasible Tr changes (i.e. the maximum stretch for a pair of vertices may change), hence potential edges that participate in its new fitness of the resultant Tr may change for the next round of an edge-deletion and edge-addition. This perturbation strategy causes the current solution to move at a distance of two from its current position in the search space.

To understand this perturbation strategy, we present a series of figures (see Figure 3.3 (a-f)) that illustrate the perturbation strategy.

Fig. 3.3(a) presents a $G = (V, E, w)$ which is connected, undirected and edge-weighted graph $G = (V, E, w)$ with $|V| = 8$, $|E| = 15$, and each edge $e \in E$ having its own positive weight (w) and a solution Tr (spanning tree) of G is shown in Figure 3.3(b). The pair of vertices between 2 and 7 has the maximum stretch in comparison to all other pairs of vertices in Tr, which contribute in the fitness of Tr, i.e. $t = 2.33$. Potential edges (highlighted in dark color) between the vertex 2 and the vertex 7 are (<2-5>, <5-6>, <6-7>), as these edges contribute for the maximum stretch of the vertex 2 and the vertex 7. The perturbation strategy, that follows the first round of an edge-deletion and edge-addition, first deletes a random potential edge (say <2-5> (shown in dashed line in Figure 3.3(c)) that is selected from potential edges (i.e. <2-5>, <5-6>, <6-7>). This divides Tr into two components, making Tr infeasible as seen in Figure 3.3(c). The edges of G are searched for a new candidate edge (different from <2-5>) whose addition to Tr can make a new feasible solution (Tr'). <3-4> is the new edge that allows Tr to a new feasible Tr'. The fitness of this new tree is 6.33 which is the maximum stretch for the pair of vertices (2-7) and its potential edges (i.e. <2-3>, <3-4>, <4-7>) are highlighted in dark color (see Figure 3.3(d)). Another round of an edge-deletion and edge-addition similar to the first round is followed both of which are shown in Figure 3.3(e)-(f). The fitness of the resultant solution (say Tr') shown in

Figure 3.3(f) is 1.5.

3.2.5 Local search improvement strategy

The perturbed solution (Tr') due to applying perturbation strategy on its starting solution (Tr) moves at a distance of two from the current position of Tr in the search space. To intensify the local region of Tr', a local search improvement strategy is applied. Our proposed local search improvement strategy (say $LS\_ImprovementStrategy(Tr')$ procedure) greedily iterates over all potential edges of Tr' one-by-one in the order in an attempt to improve its solution quality. This procedure starts by deleting first potential edge of Tr' in the order, resulting Tr' into an infeasible solution, as this partitioning led Tr' into two components. To make Tr' feasible, an edge (different from deleted edge) whose cost must be minimum among all candidate edges in E as well as must be less than the cost of deleted edge is searched. If the search is successful, then this edge is immediately added to join these two components, and the resultant tree (say Tr^*) becomes feasible, where a candidate edge means an edge that can join these two components. If the search is not successful, then the deleted edge is restored to Tr', and the local search procedure is continued for the next potential edge of Tr' in the order. Once the search is successful, the resultant Tr^* gets updated, the local search procedure starts a fresh by iterating over all potential edges of the resultant Tr' one-by-one in the order in an attempt to improve its solution quality, as potential edges of Tr^* may change from the potential edges of Tr' due to change in the fitness (improved fitness) of Tr^* as well as its constituent edges from that of Tr'.

Note that this local search procedure stops when either of two conditions holds true, where the conditions are (i) when there is no improvement after iterating over all potential edges of the current solution one-by-one in the order, and (ii) when there are at most *para_ls* times improvements in terms of the overall fitness of the solution, where *para_ls* is a parameter.

3.2.6 Acceptance criterion

As ILS performs a walk over the search space between the starting solution Tr and Tr^* based on a mechanism that combines the perturbation strategy and the local search improvement strategy, it is the role of acceptance criterion procedure (say Acceptance_Criterion(Tr, Tr^*)), that plays a vital role in balancing a trade-off between search-related intensification and diversification, in deciding whether to accept or not Tr^* as the new starting solution in the next iteration. Our proposed acceptance criterion procedure follows two strategies in this direction. The first strategy accepts Tr^* as the new starting solution in the next iteration of ILS if Tr^* is found to be more fit than Tr; otherwise Tr is continued as the new starting solution in the next iteration of ILS. The second strategy is followed only when ILS is unable to find a better solution than the best-so-far solution obtained in the previous *para_it* iterations in the search space, where *para_it* is a parameter. Such scenario shows that ILS appears to be somehow trapped in deep local optimum. To come out from this trap, the second step applies a local optimum escaping strategy based on applying a large number of edge-deletion and edge-addition in Tr, where the large number is based on $max(2, \lfloor(|V|\text{-}1) \times para\_ac\rfloor)$ and *para_ac* is a parameter. $max(a, b)$ is a function that returns a maximum value between a and b. In this strategy, the edges of Tr for the deletion is picked randomly, where the deleted edge may or may not be potential edge of Tr. A local search improvement strategy (see 3.2.5) is applied on the resultant solution. Such large number of edge-deletion and edge-addition in Tr followed by local search improvement strategy cause a large jump in the search space and leads to a new solution which acts now as a new starting solution (say Tr) for the next round of iteration of ILS for the Tr_t-SP.

It can be seen that this local optimum escaping strategy differs on two counts from the perturbation strategy (see Section 3.2.4).

3.3 Variable neighborhood Search for the T_t-SP

This section details a Basic variable neighborhood search (VNS) approach for the Tr_t-SP.

The proposed VNS framework for the T_t-SP is presented in Algorithm 10.

Algorithm 10: The proposed VNS framework tailored to the *T_t*-SP

```
Initialize with generating an initial solution T;
Best ← T;
while Termination criterion is not fulfilled do
    for k ← 1 to K_max do
        // Generate a neighborhood solution T' in the k^th neighborhood
        structure N_k of the current solution T;
        T' ← Neighborhood_Operator(T, N_k);
        T'' ← Local_Search(T');
        // Acceptance Criterion;
        if F(T) > F(T'') then
            T ← T'';
            k ← 1;
            if T is better than Best then
                Best ← T;
        else
            k ← k + 1;
    Apply Re-start Criterion;
```

Prior to applying VNS to the *T_t*-SP, the shortest paths between all vertex pairs in the input graph are precomputed. Subsequently, the following subsections explain the key components of the proposed VNS framework for the *Tr_t*-SP.

3.3.1 Solution Encoding

Similar to previous studies [92], we have adopted edge-set encoding to represent the solution. A solution in this approach consists of a selection of $|V|$-1 edges that together construct a valid spanning tree.

3.3.2 Initial Solution

To generate an initial solution to the *T_t*-SP, we have used a modification of Prim's algorithm [93]. In this modified version, rather than always choosing the edge with the lowest cost in each iteration, a random edge is selected that connects a vertex within the existing partial tree to another vertex outside of it.

Once the initial spanning tree is constructed, a local search improvement strategy is used to enhance its solution quality, providing a better starting point in the search space.

3.3.3 Computation of the fitness of a solution

The fitness of a feasible solution (say T) is computed as per the description mentioned in Section 3.2.3.

Note that the edges that participate in the fitness of the solution (spanning tree, i.e. T) are called *potential* edges of T [36].

3.3.4 Neighborhood Structures

For the T_t-SP, the neighborhood structures in VNS are problem-specific and help in finding the set of possible solutions accessible from the current one. By systematically transitioning between different neighborhood structures, VNS explores different regions of the search space, helping in identifying solutions of high quality.

We propose a set of neighborhood operators, i.e. N_k for $k = 1, \ldots, k_{max}$ defining the neighborhood structures and the order of their examination, where $k_{max} = 3$. In each neighborhood operator N_k, based on the value of k, k-times edge-deletion and edge-addition operation (*move*) is performed on the current solution T. An edge-deletion means a potential edge of T (see Section 3.2.3 about the potential edge) is randomly selected and deleted from T, partitioning T into two components and making T infeasible. Here, a potential edge of a given solution is an edge that participates in its fitness. To ensure the feasibility of T, a randomly selected edge from the set of candidate edges in G is added. Here, a candidate edge in G refers to an edge capable of linking the two disconnected components of T.

To further enhance the quality of the obtained solution T', a local search is applied (refer to Section 3.3.5). If the locally improved solution T'' proves to be superior to the current solution T', the search process resets to the first neighborhood structure, N_1, following the predefined order. In this case, T'' is adopted as the new incumbent solution, replacing T in the procedure.

3.3.5 Local Search

Once a new candidate solution is generated by exploring a neighborhood, a local search improvement strategy [95] is applied to further improve the quality of the solution. This local search systematically examines each *potential* edge of T' in sequence, aiming to enhance the solution quality. The process begins by removing the first edge (e) from T', which results in an infeasible solution as it splits T' into two disconnected components. To restore feasibility, an alternative edge (e') that can reconnect these components is searched. This edge must have the lowest possible cost among all eligible edges in E and should be less expensive than the removed edge e. If no such edge is found, e is reinstated into T', and the procedure moves on to evaluate the next edge in the sequence. Conversely, if a suitable edge e' is identified, it is incorporated into the tree, forming a new structure denoted as T''. If T'' yields a worse fitness value than T', the modification is undone by replacing e' with the original edge e, and the procedure continues with the next edge. However, if T'' exhibits an improved fitness, it replaces T', and the local search process restarts from the beginning, iterating through all edges of the updated T'. This restart is necessary because the changes in T'', including improved fitness and altered edge composition, may influence subsequent iterations.

The local search process terminates when either of the following two conditions is met: (i) after examining all potential edges of the current solution in sequence, no further improvement is observed, or (ii) the number of improvements in the overall fitness of the solution is P_ls, which is a parameter.

3.3.6 Re-start Criterion

If the solution quality remains unchanged for a predefined number of iterations (denoted as P_ReS), the algorithm resets by introducing a new initial solution (refer to Section 3.3.2). Restarting the search from a different starting point enhances the likelihood of finding better-quality solutions.

3.4 Computational Results

This section discusses on the performance evaluation of ILS and VNS for the Tr_t-SP. On a set of benchmark instances, an extensive empirical comparisons of ILS and VNS against the existing best-so-far performing ABC algorithm for the Tr_t-SP is carried out.

3.4.1 Benchmark instances

Our proposed ILS and VNS are both evaluated on the same benchmark instances used by existing ABC algorithm for the Tr_t-SP problem. These benchmark instances consist of four distinct sets of Euclidean graph instances with vertex sizes of 50, 100, 150, and 200, randomly generated within a 100×100 plane. Each Euclidean graph instance, denoted as G_x (V_x, E_x), is further transformed into three sparse graph variants, characterized by differing edge densities: $E_{x_1} = 0.8 \times E_x$, $E_{x_2} = 0.6 \times E_x$, and $E_{x_3} = 0.4 \times E_x$, where these modifications represent the deletion of 20%, 40%, and 60% of the edges from the original graph, respectively. Each instance is uniquely labeled using the notation $Q1\_Q2\_Q3$, where $Q1$ refers to the number of vertices, $Q2$ indicates the proportion of edges retained, and $Q3$ serves as an index for different instances of the same graph size and edge density. For example, the instance 100_0.0_1 represents a complete graph with 100 vertices and no edge deletion, while instances like 100_0.2_1, 100_0.4_1, and 100_0.6_1 correspond to the same 100-vertex graph but with varying edge deletions. In total, 48 unique Euclidean graph instances are used in the experiments, as reported in [36].

3.4.2 Experimental setup

The C programming language is used to code ILS and VNS for the Tr_t-SP. A Linux-based PC with a 3.2 GHz × 4 Intel Core *i5* processor and 4 GB RAM is used to execute ILS and VNS for results on a set of benchmark instances. In the literature, there are genetic algorithm [88], ant colony optimization approach [89], steady-state-genetic algorithm [90], and ABC algorithm [36] for the Tr_t-SP. From the results reported in their studies, only ABC algorithm is the current best approach for this problem; hence, we use it as the main reference approach for our comparison purpose. For each instance, ABC algorithm was executed 10 times on the same computer as the proposed ILS and VNS approaches, ILS and VNS are also executed 10 times on each instance. The termination criterion for both our proposed approaches (ILS and VNS) is to allow them to run until the best-so-far solution returned by the approach stops improving over Max_Gen successive iterations. Here, if $|V| \leq 50$, then Max_Gen is 15,000; otherwise,

Max_Gen is 5,000.

3.4.3 Parameter tuning

The proposed ILS approach uses four key parameters (*para_pert*, *para_ls*, *para_ac*, and *para_it*), while VNS uses only two parameters (P_ls, and P_ReS), whose tuning influences the quality of solutions. To identify suitable values for these parameters, we employed the automated parameter tuning tool, Irace [96]. Eight instances were randomly chosen from different groups to serve as training instances. Following the tuning procedure, we derived the parameter configurations for both ILS and VNS, outlined in Tables 3.2 and 3.3, respectively. Tables 3.2-3.3 displays the outcomes of parameter tuning, encompassing the candidate values considered, and the best configuration for each parameter related to ILS and VNS determined by Irace. These configurations serve as the default parameter setup for our approaches, remaining consistent across all experiments conducted in this chapter.

Table 3.2: Parameter Tuning for ILS

Parameters	*Candidate Values*	*Best Configuration for ILS*
para_pert	{1,**2**,3,4}	2
para_ls	{1,2,3,**4**,5}	4
para_ac	{0.01,**0.05**,0.1,0.2}	0.05
para_it	{500,1000,**2000**,3000,4000,5000}	2000

Table 3.3: Parameter Tuning for VNS

Parameters	*Candidate Values*	*Best Configuration for VNS*
P_ls	{1,2,**3**,4,5}	3
P_ReS	{500,**1000**,2000,3000,4000,5000}	1000

3.4.4 Comparison of ILS and VNS with the best-so-far approach

This section discusses the performance evaluation of the proposed ILS and VNS in comparison to existing ABC algorithm. We have not considered other existing metaheuristic techniques in the literature because their results are dominated by ABC algorithm.

Table 3.4 summarizes the empirical results of 48 Euclidean graph instances obtained by ABC algorithm, ILS and VNS. In Table 3.4, the first column presents instance

(*Instance*), the next two columns respectively present $|V|$ and the edge density (E-D) for each considered instance, and each subsequent four columns are related to results (best value (*Best*), average solution quality (*Avg*), and standard deviation (*StD*), average total time (*ATT*) over 10 runs) returned by ABC algorithm, ILS and VNS, respectively.

Based on the overall results across 48 instances reported in Table 3.4, the following observations can be made:

- ILS vs. ABC algorithm: In terms of *Best* values, ILS performs better on 32 instances, equal on 9, and worse on 7. For *Avg* values, ILS is better on 26, equal on 1, and worse on 21 instances.
- VNS vs. ABC algorithm: For *Best* values, VNS performs better on 29 instances, equal on 11, and worse on 8. In terms of *Avg* values, VNS is better on 29 instances, equal on 4, and worse on 15.

From the results reported in Table 3.4, one observes that the performance of ILS and VNS improves as the size of the instance increases. ILS is able to discover new values on 32 instances compared to ABC algorithm, whereas VNS is able to discover new values on 29 instances compared to ABC algorithm. In terms of average solution quality, ILS and VNS are more robust than ABC algorithm. It also shows that ILS and VNS dominate ABC algorithm, particularly with the increase of instance sizes.

Between VNS and ILS, VNS performs better on 13 instances, equal on 10, and worse on 25 instances in terms of finding the *Best* value. In terms of the *Avg* value, VNS is better on 23 instances, equal on 3, and worse on 22 instances compared to ILS.

Table 3.4: Computational results of ILS and VNS against existing ABC algorithm

Instances	*Characteristics*		*ABC*				*ILS*				*VNS*			
	|V|	E-D	*Best*	*Avg*	*StD*	*ATT*	*Best*	*Avg*	*StD*	*ATT*	*Best*	*Avg*	*StD*	*ATT*
50_0.0_1	50	0%	**4.67**	4.84	0.09	4.26	4.73	4.85	0.13	4.68	4.72	**4.78**	0.05	5.54
50_0.2_1	50	20%	4.33	**4.36**	0.03	4.28	4.33	4.40	0.09	4.95	4.33	4.41	0.06	4.20
50_0.4_1	50	40%	4.44	4.44	0.00	4.31	4.44	4.44	0.00	3.94	4.44	4.44	0.00	4.66
50_0.6_1	50	60%	4.64	4.66	0.04	4.28	**4.46**	4.67	0.10	4.42	4.57	**4.63**	0.04	5.16
50_0.0_2	50	0%	5.02	**5.04**	0.06	4.20	**5.00**	5.07	0.09	4.47	5.02	5.09	0.13	4.24
50_0.2_2	50	20%	4.69	4.69	0.00	4.26	4.69	4.73	0.12	4.06	4.69	4.69	0.00	4.22
50_0.4_2	50	40%	3.98	**3.98**	0.00	4.28	3.98	4.00	0.03	4.32	3.98	4.05	0.09	4.89
50_0.6_2	50	60%	4.00	**4.00**	0.00	4.23	4.00	4.03	0.03	4.37	4.00	4.02	0.03	5.75
50_0.0_3	50	0%	5.45	**5.49**	0.10	4.22	5.45	5.51	0.13	4.92	5.45	5.64	0.19	4.96
50_0.2_3	50	20%	4.68	**4.74**	0.05	4.24	4.68	4.77	0.06	4.85	4.75	4.79	0.05	4.59
50_0.4_3	50	40%	4.68	**4.74**	0.05	4.25	4.68	4.77	0.06	4.86	4.75	4.79	0.05	5.13
50_0.6_3	50	60%	4.20	4.20	0.00	4.23	4.20	4.20	0.00	2.37	4.20	4.20	0.00	3.12
100_0.0_1	100	0%	**6.17**	**6.36**	0.12	13.51	6.30	7.03	0.46	8.39	6.50	6.85	0.29	6.81
100_0.2_1	100	20%	6.04	6.19	0.09	13.48	**5.99**	**6.13**	0.13	7.38	6.17	6.38	0.20	6.38
100_0.4_1	100	40%	6.16	**6.36**	0.15	13.64	**6.11**	6.51	0.31	9.60	6.16	6.52	0.38	9.59
100_0.6_1	100	60%	6.51	**6.65**	0.10	13.72	**6.40**	7.00	0.29	7.11	6.72	6.93	0.16	9.26
100_0.0_2	100	0%	5.89	**6.12**	0.15	13.56	5.98	6.31	0.23	8.30	**5.73**	6.41	0.28	8.50
100_0.2_2	100	20%	5.93	**6.15**	0.19	13.63	6.07	6.46	0.28	7.39	**5.87**	6.34	0.27	7.81
100_0.4_2	100	40%	5.86	**6.1**	0.14	13.62	6.19	6.37	0.13	8.68	**5.83**	6.16	0.17	8.72
100_0.6_2	100	60%	5.83	6.16	0.2	13.72	5.83	6.17	0.26	5.59	5.83	**5.99**	0.18	7.40
100_0.0_3	100	0%	7.09	7.49	0.24	13.55	**6.99**	7.55	0.34	7.36	7.18	**7.38**	0.20	9.43
100_0.2_3	100	20%	7.08	7.57	0.26	13.58	7.07	7.60	0.44	8.10	**7.06**	**7.36**	0.37	8.87
100_0.4_3	100	40%	6.84	7.30	0.30	13.71	6.70	7.09	0.31	9.58	6.70	**6.98**	0.22	8.97
100_0.6_3	100	60%	6.83	7.25	0.17	13.72	**6.72**	7.17	0.24	10.01	6.83	**7.16**	0.18	7.74
150_0.0_1	50	0%	8.12	9.06	0.44	28.33	7.99	8.92	0.61	15.79	**7.96**	**8.80**	0.45	19.89
150_0.2_1	50	20%	8.67	9.13	0.37	28.67	7.82	8.66	0.59	19.15	7.82	**8.25**	0.28	23.44
150_0.4_1	50	40%	7.54	8.41	0.45	29.03	7.33	7.76	0.34	20.68	**7.13**	**7.44**	0.14	17.08
150_0.6_1	50	60%	6.86	7.57	0.49	29.02	**6.20**	6.75	0.35	18.34	6.43	**6.66**	0.15	20.84

Table 3.4 Continued: Computational results of ILS and VNS against existing ABC algorithm

Instances	*Characteristics*		*ABC*				*ILS*				VNS			
	|V|	*E-D*	*Best*	*Avg*	*StD*	*ATT*	*Best*	*Avg*	*StD*	*ATT*	*Best*	*Avg*	*StD*	*ATT*
150_0.0_2	50	0%	8.97	9.40	0.28	28.56	**8.69**	9.54	0.56	16.95	8.84	**9.23**	0.28	21.33
150_0.2_2	50	20%	8.53	8.95	0.27	28.62	**8.2**	**8.68**	0.22	23.17	8.46	8.76	0.22	21.17
150_0.4_2	50	40%	**8.00**	8.78	0.52	28.96	8.10	**8.66**	0.36	20.42	8.27	8.75	0.34	19.64
150_0.6_2	50	60%	7.28	8.20	0.49	28.88	6.84	7.48	0.26	15.59	**6.72**	**7.13**	0.24	25.24
150_0.0_3	50	0%	11.01	11.53	0.42	28.41	9.98	**10.60**	0.50	22.68	**9.71**	10.63	0.52	22.98
150_0.2_3	50	20%	10.39	10.84	0.42	28.66	9.59	**10.03**	0.36	29.39	**9.57**	10.11	0.55	25.85
150_0.4_3	50	40%	10.00	10.72	0.69	28.90	**8.25**	**9.45**	0.92	25.43	8.36	9.56	0.71	23.84
150_0.6_3	50	60%	8.63	9.31	0.48	29.16	**7.24**	**7.84**	0.39	23.43	7.51	7.86	0.24	26.22
200_0.0_1	100	0%	9.96	11.63	0.72	49.63	**9.71**	**10.37**	0.35	43.46	10.07	10.57	0.61	35.17
200_0.2_1	100	20%	11.20	12.28	0.76	50.68	9.67	10.28	0.37	43.93	**9.15**	**9.79**	0.36	40.26
200_0.4_1	100	40%	10.43	11.71	0.61	52.55	**8.53**	**9.53**	0.55	42.08	8.97	9.80	0.37	35.42
200_0.6_1	100	60%	9.57	10.53	0.64	50.67	**8.11**	**8.83**	0.44	36.7	8.59	9.15	0.39	34.79
200_0.0_2	100	0%	11.57	13.41	1.13	50.11	11.23	11.89	0.53	37.51	**11.08**	11.89	0.60	36.16
200_0.2_2	100	20%	11.06	12.09	0.68	49.56	10.39	11.1	0.35	41.38	**10.23**	**10.97**	0.52	32.82
200_0.4_2	100	40%	11.17	11.74	0.45	50.59	**9.85**	**10.51**	0.34	52.14	10.22	11.13	0.96	30.60
200_0.6_2	100	60%	10.14	11.16	0.70	50.85	8.99	9.61	0.63	42.25	**8.64**	**9.33**	0.25	45.12
200_0.0_3	100	0%	12.37	13.35	0.63	49.66	**10.86**	**11.65**	0.53	36.96	11.02	11.80	0.43	38.17
200_0.2_3	100	20%	12.47	14.04	1.18	49.70	**11.13**	**12.08**	1.72	50.70	11.23	12.17	0.52	37.91
200_0.4_3	100	40%	11.77	13.27	1.01	50.67	**9.65**	13.60	2.52	50.64	11.76	**12.61**	0.69	44.42
200_0.6_3	100	60%	10.90	12.07	1.13	50.54	**9.32**	**10.36**	0.82	50.41	9.58	10.43	0.76	52.85
Total						1164.89				928.88				887.15

From a computational time perspective, the final row of Table 3.4 displays the average of the average total time (ATT) across all instances, i.e. *Total* for ABC algorithm, ILS, and VNS. Specifically, the *Total* for the ABC algorithm, ILS, and VNS are 1164.89 seconds, 928.88 seconds, and 887.15 seconds respectively. Overall, these results demonstrate that ILS and VNS are found to be computationally faster than ABC algorithm on considered instances for the *Tr_t*-SP. Between VNS and ILS, VNS is computationally faster than ILS on the considered instances for the *T_t*-SP problem.

3.4.5 Analysis of the use of the second strategy in the acceptance criterion of ILS

To understand the impact of the use of the second strategy in the accceptance criterion of ILS (see Section 3.2.6), we have performed an empirical analysis. To perform this analysis, the use of the second strategy in the acceptance criterion of ILS is disabled and call this approach ILS-2ndStrategy. Note that the first step of acceptance criterion is not disabled from ILS-2ndStrategy. We use all considered instances for our analysis and report the results (*Best* and *Avg* over 10 runs) of these instances obtained by ILS and ILS-2ndStrategy (see Table 3.5). We also report *%Gap-Best* value ($\frac{(Best^{ILS\text{-}2ndStrategy} - Best^{ILS})}{Best^{ILS}} \times 100\%$) and *%Gap-Avg* value ($\frac{(Avg^{ILS\text{-}2ndStrategy} - Avg^{ILS})}{Avg^{ILS}} \times 100\%$) in Table 3.5 for each instance, where $Best^{ILS\text{-}2ndStrategy}$ and $Best^{ILS}$ present the best value return by ILS-2ndStrategy and ILS; and $Avg^{ILS\text{-}2ndStrategy}$ and Avg^{ILS} present the average value over 10 runs return by ILS-2ndStrategy and ILS. *Best* and *Avg* values are highlighted in bold. One can notice from the averaging the gap values between ILS-2ndStrategy and ILS across all instances, both the average gap values (*AVERAGE*) for *%Gap-Best* and *%Gap-Avg* are positive for both, suggesting that ILS outperforms ILS-2ndStrategy on average. This analysis suggests that ILS benefits from the use of the second strategy in the acceptance criterion of ILS, and hence benefits from escaping out from being trapped in deep local optimum.

Table 3.5: Computational results of ILS-2ndStrategy and ILS on graph instances with different sizes and densities

Instances	Characteristics		ILS-2ndStrategy		ILS			
	|V|	*E-D*	*Best*	*Avg*	*Best*	*Avg*	*%Gap-Best*	*%Gap-Avg*
50_0.0_1	50	0%	4.79	5.01	**4.73**	**4.85**	1.27	3.3
50_0.2_1	50	20%	4.33	4.51	4.33	**4.4**	0	2.5
50_0.4_1	50	40%	4.44	4.49	4.44	**4.44**	0	1.13
50_0.6_1	50	60%	4.64	4.79	**4.46**	**4.67**	4.04	2.57
50_0.0_2	50	0%	5.02	5.26	**5**	**5.07**	0.4	3.75
50_0.2_2	50	20%	4.69	4.86	4.69	**4.73**	0	2.75
50_0.4_2	50	40%	3.98	4.23	3.98	**4**	0	5.75
50_0.6_2	50	60%	4	4.13	4	**4.03**	0	2.48
50_0.0_3	50	0%	5.45	5.83	5.45	**5.51**	0	5.81
50_0.2_3	50	20%	4.75	4.94	**4.68**	**4.77**	1.5	3.56
50_0.4_3	50	40%	4.75	4.94	**4.68**	**4.77**	1.5	3.56
50_0.6_3	50	60%	4.2	4.36	4.2	**4.2**	0	3.81
100_0.0_1	100	0%	6.41	7.3	**6.3**	**7.03**	1.75	3.84
100_0.2_1	100	20%	6.04	6.43	**5.99**	**6.13**	0.83	4.89
100_0.4_1	100	40%	6.29	6.64	**6.11**	**6.51**	2.95	2
100_0.6_1	100	60%	6.65	7.04	**6.4**	**7**	3.91	0.57
100_0.0_2	100	0%	**5.82**	6.46	5.98	**6.31**	-2.68	2.38
100_0.2_2	100	20%	**5.93**	**6.45**	6.07	6.46	-2.31	-0.15
100_0.4_2	100	40%	**5.53**	6.48	6.19	**6.37**	-10.66	1.73
100_0.6_2	100	60%	5.83	**6.15**	5.83	6.17	0	-0.32
100_0.0_3	100	0%	7.12	7.61	**6.99**	**7.55**	1.86	0.79
100_0.2_3	100	20%	**7.01**	7.72	7.07	**7.6**	-0.85	1.58
100_0.4_3	100	40%	6.7	7.09	6.7	7.09	0	0
100_0.6_3	100	60%	7.04	7.92	**6.72**	**7.17**	4.76	10.46
150_0.0_1	50	0%	8.32	9.3	**7.99**	**8.92**	4.13	4.26
150_0.2_1	50	20%	**7.69**	8.92	7.82	**8.66**	-1.66	3
150_0.4_1	50	40%	**6.97**	7.98	7.33	**7.76**	-4.91	2.84
150_0.6_1	50	60%	6.49	7.13	**6.2**	**6.75**	4.68	5.63

Table 3.5 Continued: Computational results of ILS-2ndStrategy and ILS on graph instances with different sizes and densities

Instances	**Characteristics**		**ILS-2ndStrategy**		**ILS**			
	|V|	*E-D*	*Best*	*Avg*	*Best*	*Avg*	*%Gap-Best*	*%Gap-Avg*
150_0.0_2	50	0%	8.55	**9.36**	**8.69**	9.54	-1.61	-1.89
150_0.2_2	50	20%	**7.87**	9.25	8.2	**868**	-4.02	6.57
150_0.4_2	50	40%	**7.94**	**8.57**	8.1	8.66	-1.98	-1.04
150_0.6_2	50	60%	6.85	7.27	**6.84**	**7.48**	0.15	-2.81
150_0.0_3	50	0%	10.2	11.43	**9.98**	**10.6**	2.2	7.83
150_0.2_3	50	20%	9.22	10.44	**9.59**	**10.03**	-3.86	4.09
150_0.4_3	50	40%	**8.2**	10.2	8.25	**9.45**	-0.61	7.94
150_0.6_3	50	60%	7.4	7.84	**7.24**	7.84	2.21	0
200_0.0_1	100	0%	10.06	10.69	**9.71**	**10.37**	3.6	3.09
200_0.2_1	100	20%	10.19	11.29	**9.67**	**10.28**	5.38	9.82
200_0.4_1	100	40%	9.12	9.72	**8.53**	**9.53**	6.92	1.99
200_0.6_1	100	60%	8.42	9.44	**8.11**	**8.83**	3.82	6.91
200_0.0_2	100	0%	**10.91**	12.41	11.23	**11.89**	-2.85	4.37
200_0.2_2	100	20%	**10.1**	**10.63**	10.39	11.1	-2.79	-4.23
200_0.4_2	100	40%	10.14	12.3	**9.85**	**10.51**	2.94	17.03
200_0.6_2	100	60%	**7.49**	9.93	8.99	**9.61**	-16.69	3.33
200_0.0_3	100	0%	11.08	12.49	**10.86**	**11.65**	2.03	7.21
200_0.2_3	100	20%	**11.01**	13.18	11.13	**12.08**	-1.08	9.11
200_0.4_3	100	40%	11.83	14.01	**9.65**	**13.6**	22.59	3.01
200_0.6_3	100	60%	**8.86**	10.97	9.32	**10.36**	-4.94	5.89
AVERAGE							0.46	3.6

3.4.6 Statistical Analysis

Table 3.6: Statistical test on the results of *Best* values

Approches	p-value	significant
ILS vs ABC	0.00001	yes
VNS vs ABC	0.00022	yes

Table 3.7: Statistical test on the results of *Avg* values

Approches	p-value	significant
ILS vs ABC	0.00362	yes
VNS vs ABC	0.00001	yes

A statistical analysis was conducted to compare the results of ILS, VNS, and existing ABC algorithm for the Tr_t-SP across all instances. The analysis utilized a Wilcoxon's Signed-Rank test calculator [69] to perform a two-tailed non-parametric Wilcoxon's signed rank test, with a significance criterion set at 0.05. The statistical results of ILS vs ABC algorithm and VNS vs ABC algorithm presented in Tables 3.6 and 3.7 for *Best* and *Avg* values, respectively. It was found that ILS and VNS exhibited significant differences compared to ABC algorithm in terms of *Best* values across all instance sizes. Regarding *Avg* values, the statistical analysis also revealed significant differences between ILS and VNS with ABC algorithm.

Chapter 4

Single Machine Scheduling Problem With Periodic Machine Availability

4.1 Introduction

In production, deterministic maintenance (maintenance of machine parts) or an uncertain phenomenon (failure of machine parts) are often required, causing unavailability of machines continuously to process all jobs. These stoppages directly affect overall productivity efficiency. Consequently, the industry has increasingly prioritized maintenance activities within their production systems. Scheduled maintenance involves regular inspection and upkeep of machine parts to maintain peak performance and prevent minor issues from escalating into major problems. Through careful production scheduling and regular maintenance planning, not only can production efficiency and safety be enhanced, but productivity can also be increased. This study addresses a scheduling problem concerning a single machine with periodic machine availability with the objective of minimizing the makespan of job scheduling. Henceforth, this problem will be referred to as the PMASMSP.

The PMASMSP comprises a set of n jobs (named $\{J_1, J_2, \ldots, J_n\}$). Each job is associated with a processing time p_j, where $1 \leq j \leq n$. Each job becomes available at time zero. The schedule divides the planning horizon into periods $(P_1, P_2, \ldots, P_L)$, where each period is of T time unit. During these periods, machine availability, referred to as periodic machine availability, alternates with other periods (referred to as maintenance activities or non-availability periods). These periods endure for t time units, during which no operation is performed (see Figure 4.1).

commenting for now

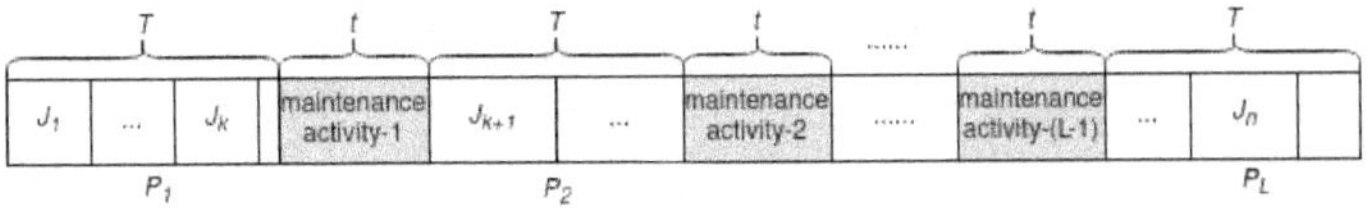

Figure 4.1: PMASMSP

To ensure the feasibility of the PMASMSP, it is required that $T \geq p_j \, \forall j = \{1, 2, \ldots, n\}$. This is termed as $1|nr\text{-}pm|C_{max}$ by Ji *et al.* [97], where "1" signifies a single machine; "*nr-pm*" denotes a constraint regarding non-resumable (*nr*) jobs during periodical maintenance (*pm*) non-availability periods, implying that if a job fails to complete processing before maintenance, it must restart. Finally, "C_{max}" indicates the objective of minimizing the makespan of the schedule.

Such problem is referred as scheduling with periodic maintenance [97, 98], and when such problem considers availability periods constraint, then it becomes $\mathcal{NP}$-hard [99].

4.1.1 Literature survey

The literature has been several prior works on problems related to the PMASMSP. Hsu *et al.* [100] worked on a variant of the PMASMSP where machine operation is interrupted for maintenance either after processing a certain number (say K) of jobs or after a fixed time interval of T units. For this variant problem, Hsu *et al.* worked on two approaches. The first one is a Binary Integer Programming (BIP) model structured in two stages, while the second approach involves two heuristics in which the first heuristic is decreasing order with the best-fit bin packing strategy and the second one is Butterfly order with the best-fit bin packing strategy. Later, a variant incorporating flexible (not fixed) periodic maintenance intervals within a given time window was studied by Low *et al.* [101]. They also constrained job assignments to availability periods, akin to Hsu *et al.*'s K job constraint [100]. Low *et al.* [101] continued to work with this variant and presented six constructive heuristics adopting the idea of combining a sequencing priority list (longest processing time (LPT) order, shortest processing time order, and random order) with the first-fit/best-fit strategies. Later, Shen and Zhu [102] worked on a single machine scheduling problem with periodic maintenance and uncertain processing time.

For the PMASMSP, several constructive heuristics have been studied by various researchers. Ji *et al.* [97] utilized the concept of LPT along with a first-fit bin packing policy, exhibiting a worst-case ratio of 2, to propose a constructive heuristic. Yu *et al.* [98] introduced two constructive heuristics, both with a worst-case bound of 2. The first heuristic is based on the list scheduling algorithm employing the first-fit bin packing strategy, while the second heuristic is the MLPT (modified LPT) that draws inspiration from the bin packing problem [103]. Low *et al.* [104] experimented with four variants of particle swarm optimization (PSO) (i.e. PSO-M, PSO-R,PSO-1 and PSO-2) for the PMASMSP. Among these four variants, PSO-M outperforms the other three variants of PSO. A detail of four variants of PSO can be studied in [104]. Later, Paz and Jose [37] carried out a computational evaluation of 15 constructive heuristics on the PMASMSP, where these 15 constructive heuristics are taken from different papers ([97],[98],[104],[100],[101]). These 15 existing constructive heuristics consist of two phases. In the first phase, certain sorting criterion depending on the processing times of the jobs are used to sort the jobs, whereas, the second phase employes a specific bin packing strategy. Table 4.1 presents these 15 constructive heuristics. Readers are referred to study all cited papers in the paper [37] in detail. Moreover, Paz and Jose [37] also presented 3 new heuristics that use bin packing assignment policy as an operator and termed them as NEW, NEW_FF and NEW_BF heuristics depending on the operator none, first-fit or best-fit policy applied. After that a simple local search is applied on these heuristics to explore the space of solutions. It was demonstrated that NEW_BF and NEW_FF perform superior. Also, Paz and Jose [37] introduced two mixed integer linear programming (MILP) models. The first model called BP-MILP was originally developed for the bin packing problem [105]. To adapt this model for the PMASMSP, Paz and Jose [37] used the notation $1|nr\text{-}pm|C_{max}$ that allots the jobs to the periods minimizing the number of periods. PM-MILP, a second model, is an adaptation of the BP-MILP model for the problem $1|nr\text{-}pm|C_{max}$. It was observed that the optimal solution from PM-MILP also exists within the optimal solutions generated by BP-MILP, specifically involving the block with the highest slack value. The optimal solutions derived from PM-MILP form a subset of those produced by BP-MILP. Moreover, they also pointed that although both problems, namely the bin packing problem and the PMASMSP, differ in practice, the solutions yielded by both models may exhibit significant similarity. Thus, a total of 19 approaches was studied by Paz and Jose [37]. Recently, a steady-state grouping genetic algorithm (say GA) [38] was proposed for the PMASMSP. Authors [38] showed that their proposed GA outperforms all 19 existing approaches.

Table 4.1: All 15 Constructive Heuristics

Methods	Reference(s)
First Fit Random (FFR)	Yu *et al.*[98], Low *et al.* [101]
First Fit Decreasing (FFD)	Ji *et al.* [97], Low *et al.* [104], Low *et al.* [101], Yu *et al.* [98]
First Fit Increasing (FFI)	Low *et al.* [104], Low *et al.* [101]
First Fit V-Sharp (FFV)	Low *et al.* [104]
First Fit A-Sharp (FFA)	Low *et al.* [104]
First Fit HILO (FFHILO)	Low *et al.* [104]
First Fit LOHI (FFLOHI)	Paz and Jose [37]
Best Fit Random (BFR)	Low *et al.* [101]
Best Fit Decreasing (BFD)	Hsu *et al.* [100], Low*et al.* [104], Low *et al.* [101]
Best Fit Increasing (BFI)	Low *et al.* [104], Low *et al.* [101]
Best Fit V-Sharp (BFV)	Low *et al.* [104]
Best Fit A-Sharp (BFA)	Low *et al.* [104]
Best Fit HILO (BFHILO)	Hsu *et al.* [100], Low *et al.* [104]
Best Fit LOHI (BFLOHI)	Paz and Jose [37]
Modified FFD (MFFD)	Yu *et al.* [98]

4.2 Our contribution

Despite being $\mathcal{NP}$-hard, only two metaheuristic techniques, i.e. PSO [104] and GA [38] have been studied in the literature for the PMASMSP, and both are population-based metaheuristic techniques. No single solution-based metaheuristic technique has been studied and explored for this problem. This motivated us to work on an iterated local search (ILS) approach – a well-known single solution-based metaheuristic technique – for this problem. The proposed ILS is effective and superior to the best 3 out of 19 methods [37] as well as GA [38] across the available benchmark instances in the literature. Our proposed ILS finds new best results in 6 out of the 14 categories of benchmark instances.

The next remaining sections of this chapter is framed as follows. In Section 4.3, we discuss a mathematical model for the PMASMSP. In Section 4.4, we present our proposed iterated local search for the PMASMSP. In Section 4.5, we provide computational results, comparison with previous methods, and a study on the use of 2nd step procedure in acceptance criterion of ILS.

4.3 Problem description

This section presents a mixed integer linear programming (MILP) model (referred to as the PM-MILP model), which was proposed by Paz and Jose [37] for the PMASMSP. The following notation is used for the PM-MILP model:

- □ Data
  - n (the number of jobs);
  - p_j (the processing times of jobs $1 \leq j \leq n$);
  - T (the availability period);
  - Z (a large positive number $\leq T$).
- □ Variables
  - b_i (binary variables indicating if block i is occupied, $1 \leq j \leq n$);
  - x_{ij} (binary variables indicating if job j is allotted to the block i, $1 \leq j \leq n$, $1 \leq j \leq n$);
  - *maximum_slack* (a variable having the maximum slack or empty space of the period);
  - δ_i (binary variables indicating if the maximum slack is provided by the period i);

PM-MILP model:

$$minimum\ ((T\sum_{i=1}^{n} b_i) - max\_slack) \tag{4.1}$$

$$\sum_{j=1}^{n} p_j x_{ij} \leq Tb_i, \quad 1 \leq i \leq n \tag{4.2}$$

$$\sum_{i=1}^{n} x_{ij} = 1, \quad 1 \leq i \leq n \tag{4.3}$$

$$maximum\_slack \leq Z(1-\delta_i) + Tb_i - \sum_{j=1}^{n} p_j x_{ij} = 1, \quad 1 \leq i \leq n \tag{4.4}$$

$$\sum_{j=1}^{n} x_{ij} \geq b_i, \quad 1 \leq i \leq n \tag{4.5}$$

$$\sum_{i=1}^{n} \delta_i = 1 \tag{4.6}$$

$$\delta_i \leq b_i, \quad 1 \leq i \leq n \tag{4.7}$$

$$x_{ij}, b_i, \delta_i \in \{0,1\}, \quad 1 \leq i, j \leq n \tag{4.8}$$

$$maximum\_slack \geq 0 \tag{4.9}$$

where Equation 4.1 is the objective to be minimized, i.e. the makespan which can be computed as (T × (*the total number of used periods*)) - (*the maximum slack of the periods*). Equation 4.2 imposes the constraint that the cumulative processing time of all jobs allotted to an utilized period (denoted as *i*) must not exceed T. Equation 4.3 impose that each job must be allotted to one and only one period, for all jobs. Equation 4.4 sets the maximum slack among the used periods. For each period, Equation 4.5 says that, if a period is not used, then no job is allotted to this period. Equation 4.6 says that only one period provides the maximum slack, and Equation 4.7 says that, if a period is not used, then this period does not provide the maximum slack. Finally, Equations 4.8 and 4.9 respectively define the binary and continuous variables used in this model.

Similar to Paz and Jose's approach [37], we have opted not to include the value of t (refer to 4.1) due to its lack of impact on the optimal solution.

4.4 Iterated local search for the PMASMSP

This section discusses our proposed iterated local search (ILS) for the PMASMSP.

The components of the proposed ILS for the PMASMSP are explained in subsequent subsections.

4.4.1 Solution encoding

Each feasible solution to the PMASMSP can be viewed as a collection of periods (see Figure 4.2), as all jobs are partitioned (assigned) into a set of production periods. Therefore, the encoding of each solution is also a collection of periods (i.e., $\{P_1, P_2, \ldots, P_L\}$), where each job is assigned to one of these periods. Here, L denotes the sum of all periods required for scheduling n jobs for a given instance of the problem. For example, for a given instance with $n = 5$ jobs i.e. $\{J_1, J_2, J_3, J_4, J_5\}$ along with their processing times $\{6, 7, 4, 8, 2\}$, a Gantt-chart based on its a feasible solution (say π) is illustrated in Figure 4.2. Here, a feasible solution (π) is represented as a collection of 3 periods $\{P_1, P_2, P_3\}$ for scheduling all 5 jobs, where P_1 contains J_1 and J_3 jobs; P_2 contains J_2 and J_5 jobs; and P_3 contains J_4 job. The sequence of jobs in this schedule of π is $\{J_1, J_3, J_2, J_5, J_4\}$. Each period of a solution π satisfies its time constraint T (the value of T is 11). The makespan of this schedule (π) is 30.

commenting for now

4.4.2 Initial solution generation

The initial start solution (say π) in ILS is generated randomly. The procedure for generating this initial solution (referred to as *InitialSolutionGeneration*) begins with all n jobs initially placed in a set called U. After this, the procedure, at each iteration,

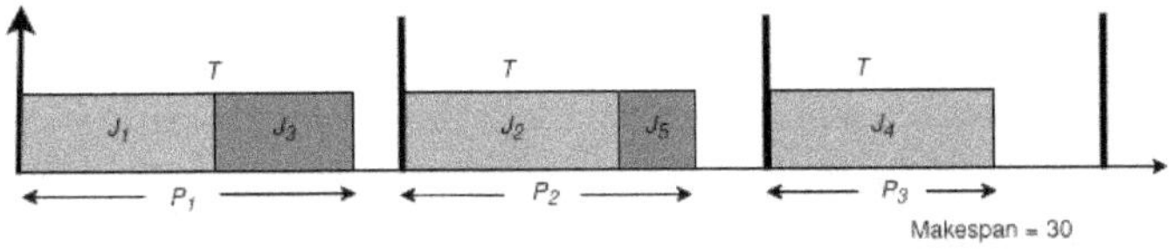

Figure 4.2: The Gantt-chart of a schedule (π)

creates a new empty period (designated as P_1). For this period, a job (denoted as J_r) from U is randomly selected. The selection of a job is contingent upon its insertion into this period without violating the constraint T; otherwise, the job is discarded. If the job J_r is picked, then J_r is deleted from the set U. This iterative process continues until all jobs in U have been considered for insertion into this period. If job(s) still exist(s) in U, then the next iteration begins with taking a new empty period (i.e. P_2) into account for inserting randomly selected jobs from U, ensuring compliance with the T constraints. This iterative procedure persists until there is no job in U, signifying that all jobs in U have been allotted to a collection of periods $\{P_1, P_2, \ldots, P_L\}$, where L denotes the sum of all periods in the solution π.

It is to be noted that the sum of all periods can differ between individual solutions.

4.4.3 Computation of the fitness of a solution

Since the fitness ($F(\pi)$) of a solution (π) is its makespan, its makespan is calculated according to *Equation*4.1. Illustrated in Figure 4.2 is a feasible solution (π) for a given instance, requiring 3 periods to schedule all jobs. Thus, the fitness of this solution (π) is calculated as (3 × 11) - (11 - 8) = 30, following *Equation*4.1 of the PM-MILP model, where 11-8 = 3 is the slack value of the last period. Its fitness ($F(\pi)$) can be also computed as follows:

$F(\pi)$ = *((the total number of periods except the last period)* ×*T*) + *(the total processing times for all jobs in the last period)*, i.e. (2 × 11) + 8 = 30.

It is to be noted here that if the *slack* value of last period of a solution increases or the sum of processing times for all jobs in the last period decreases, then fitness of this solution decreases or the quality of this solution improves.

4.4.4 Perturbation procedure

In the perturbation procedure, the current solution (π) is perturbed in such a way that the resultant perturbed solution can move from its current location to a new location in the search region. The perturbation procedure first creates a copy (π') of the current solution (π), then, a fix number (say $X = (int)(Para_{pert} \times n)$) of iteration is executed, where $Para_{pert}$ is a parameter. In each iteration, two different periods (say P_i and P_j) of π' are picked randomly. One job from each picked period is picked randomly. After this, it is checked whether picked jobs can be swapped without violating the constraint. If the constraint is not violated, then picked jobs are swapped, otherwise, the next iteration starts. This way of perturbation allows a subset of jobs to move into different periods of the current solution (π), resulting a new perturbed solution (π').

4.4.5 Local search (LS)

The aim of the local search within ILS is to exploit the search locally around the perturbed solution (π') region, aiming to find a superior solution quality. For the PMASMSP, we employ a series of local search strategies, i.e. 1-0 insertion (say $LS\_10\_Insertion$), 1-1 *swap* (say LS_11_Swap), 1-2 *swap* (say LS_12_Swap), and 2-1 *swap* (say LS_21_Swap). Here, the local search 1-0 insertion ($LS\_10\_Insertion$) means each job (in the sequence) of the last period ($P'_{L'}$) of π' is tried for insertion into one of its remaining periods (checking from first period to second last period of π') in such a way that the constraint T is not violated. Such insertion reduces the total processing time of the last period of the solution, thereby reducing overall the makespan of π' (improvement in solution quality of π'). It is to be noted that if there is only one job left in the last period of π', and if the local search 1-0 insertion gets successful in inserting this last job in one of remaining periods, then, the total number of periods of π' is reduced by one. It means that the second last period becomes the last period of π'. In such condition, $LS\_10\_Insertion$ is continued in the same manner for this new last group of π'. It is clear that this $LS\_10\_Insertion$ also attempts to reduce the number of periods of the solution (π').

The local search 1-1 *swap* (LS_11_Swap) means one job (in the sequence) of the last period ($P'_{L'}$) of π' is tried for swap with one job (in the sequence) belongs to one of the remaining periods of π' (checking from first period to second last period of π') subject

Algorithm 11: A framework of Local search strategies for the PMASMSP

Input: A perturbed solution (π')

Output: A local optimum solution (π'')

```
K ← 1;
while (K < 5) do
    switch K do
        case 1 do
            Improve ← LS_21_Swap(π');
            break;
        case 2 do
            Improve ← LS_12_Swap(π');
            break;
        case 3 do
            Improve ← LS_11_Swap(π');
            break;
        case 4 do
            Improve ← LS_10_Insertion(π');
            break;
    if (Improve) then
        K ← 1;
    else
        K ← K + 1;
return π'';
```

to the constraint T. In 1-2 *swap* (LS_12_Swap) means one job (in the sequence) of the last period ($P'_{L'}$) of π' is tried for *swap* with two consecutive jobs (in the sequence) belong to one of the remaining periods of π' (checking from first period to second last period of π') subject to the constraint T, whereas 2-1 *swap* (LS_21_Swap) means two consecutive jobs (in the sequence) of the last period ($P'_{L'}$) of π' is tried for *swap* with one job (in the sequence) belongs to one of the remaining periods of π' (checking from first period to second last period of π') subject to the constraint T. If such *swap* does not violate the constraint T, and the resultant *swap* reduces the total processing time of the last period of the solution, then it causes in the reduction of the makespan of π' (improvement in the solution quality of π'), and such *swap* is carried out.

It is to be noted that all local search strategies emphasize on the last period of the solution, as the fitness of the solution improves if the total processing time of the last period reduces (see the last paragraph of Section 4.4.3).

Algorithm 11 outlines how all local search strategies ($LS\_10\_Insertion$, LS_11_Swap, LS_12_Swap, and LS_21_Swap) are applied. Our initial experiments suggested that applying local search strategies in this order mentioned in Algorithm 11 returns the best results. It is to be noted that the local search strategies start from the first one again, if one local search strategy becomes successful.

4.4.6 Acceptance criterion

Once the local search is completed, the acceptance criterion which is the last component of ILS is applied. This acceptance criterion takes decision about which solution will be selected as the new starting solution in the next iteration of ILS. Our acceptance criterion follows two step procedures.

During the 1st step procedure, if the fitness of the locally optimal solution (π'') surpasses that of the current solution (π), the locally optimal solution (π'') will be chosen as the starting point for the subsequent iteration of ILS. Conversely, if the current solution (π) proves to be superior, it will persist as the starting point.

The 2nd step procedure is applied only when the 1st step procedure repeatedly carries the same current solution as the new starting solution over a subsequent number of iterations (denoted as $Para_{iter}$). If this situation occurs, it is presumed that the current solution of ILS is trapped in a deep local optima.

To prevent this occurrence, the 2nd step procedure, with a probability ($Para_{pb}$), replaces the current solution with an initial solution through the initial solution procedure (see Section 4.4.2). Alternatively, with a probability (1-$Para_{pb}$), the 2nd step procedure replaces the current solution with a solution through the *DRP* procedure which is based on the destruction of some periods of the current solution (π). After this, a repair operator is applied in order to make the solution feasible. Here, $Para_{pb}$ is a parameter.

In the *DRP* procedure, a certain number ($X = (int)(Para_{drp} * cnt)$) of non-singleton periods of the current solution (π) are destroyed along with singleton periods of π, making π infeasible. The total number of non-singleton periods in π is cnt, and $Para_{drp}$ is a parameter. Jobs belonging to all destroyed periods are placed in a set called U. To make π feasible, a repair operator that consists of two parts is proposed. In the first part of repair operator, each job in set U undergoes scrutiny for potential insertion into all existing periods of solution π without violating T constraint. A candidate job with minimum *slack* among all jobs in U is permitted to insert into that the existing period of π. Inserted job is now removed from U. This first part is continued until no other job in U is inserted into any existing period of π without violating constraint. After completing the first part, if there are remaining jobs in set U, the repair operator proceeds to its second part, which is a greedy approach. In this second part, a new empty period is created, and the job with the longest processing time among those remaining in U is selected for insertion into this newly created period. The inserted job is then removed from U. Iteratively, a job in U having longest processing time or minimum *slack* for this newly created period is inserted without violating the T constraint. If there are still jobs in U that cannot be inserted into this newly created period, the second part of the repair operator starts fresh by creating a new empty period, and the same greedy procedure is applied to insert jobs from U into this newly created period. This second part of the repair operator persists until U becomes empty. The greedy procedure aims to insert a job with the longest processing time from set U into each newly created period as much as possible, so that the last period of π will have jobs with the minimum processing times, and the total number of periods in π is minimized.

Algorithm 12 outlines the framework (pseudo-code) of ILS for the PMASMSP.

4.5 Computational results

To examine the performance of the proposed ILS, we test it on available benchmark instances and compare it with a set of 19 approaches proposed by Paz and Jose [37] (see Section 4.1 in detail) and GA [38].

Algorithm 12: The framework (pseudo-code) of ILS for the PMASMSP

Input: Given an input of the PMASMSP

Output: The best solution returned by ILS

Apply *InitialSolutionGeneration* procedure to generate a random initial solution (say π);
while *The stopping criterion is not fulfilled* **do**
    $\pi' \leftarrow Perturbation(\pi)$;
    $\pi'' \leftarrow LocalSearch(\pi')$;
    `// The 1st step procedure of the acceptance criterion`
    if *the fitness of* π'' *is better than* π **then**
        $\pi \leftarrow \pi''$;
    `// The 2nd step procedure of the acceptance criterion`
    else
        if *The same current solution* π *is continued as the new starting solution over a subsequent number of* $Para_{iter}$ *iterations* **then**
            if $u01 < Para_{pb}$ **then**
                Replace π'' with a new starting solution (say π) generated through *InitialSolutionGeneration* procedure;
            else
                Replace π'' with a new starting solution (say π) through applying *DRP* procedure and repair operator;

4.5.1 Benchmark Instances

The proposed ILS was evaluated on the same benchmark instances used in [37]. These instances are categorized into two class of benchmark instances called *LOW* and *MOD* instances.

- □ *LOW* instances: Paz and Jose [37] generated these instances by using the method given in [100, 104].
- □ *MOD* instances: Paz and Jose [37] generated these instances by modifying (tighting) $T's$ values (i.e. $T \sim U[50, 100]$) in [100, 104]).

The values of n and p_j remain same in both class of instances, where n denotes the total number of jobs and p_j denotes the processing time for each job j is $p_j \sim U[50, 100]$. Further, each class of instances has 14 categories, with n falling within the range of $\{10, 20, 30, 40, 50, 60, 70, 80, 90, 100, 150, 200, 250, 300\}$. In each category, there are also 50 instances. Hence, each class of benchmark instances comprises of 700 instances.

4.5.2 Experimental protocol

C programming language is used to implement the proposed ILS, and the experiments were conducted on a computer equipped with Linux 2.6 GHz × 12 Intel Core *i7* - 10750H processor and 8 GB RAM.

ILS is compared with 19 approaches (15 constructive heuristic (see Table 4.1), 3 heuristics (NEW, NEW_FF and NEW_BF), and a MILP model) [37] (see detail in Section 4.1) and GA [38]. Authors [37] provided all instances and its result obtained by all their 19 approaches. Paz and Jose [37] demonstrated result for each category of instances with respect to average relative percentage deviation (*ARPD*), where the relative percentage deviation (*RPD*) for each instance is calculated as follows:

$$RPD = \frac{C_{max}(METH) - MIN}{MIN} \times 100 \tag{4.10}$$

where MIN refers to the minimum makespan among all 19 approaches, where $C_{max}(METH)$ refers to the makespan value obtained by each approach. *ARPD* refers to the average RPD.

Similar to GA [38], we have also compared the results of ILS with the results of all 19 approaches [37] and demonstrated the results of ILS with respect to ARPD. The computation of the *RPD* for each instance is

$$RPD = \frac{C_{max}(METH^{ILS}) - MIN}{MIN} \times 100 \tag{4.11}$$

It is to be noted that similar to GA [38], ILS is also given for single run on each instance. For the stopping criterion of ILS, ILS stops executing when the best-so-far solution ceases to improve over a certain number (say Max_Iter) of generations, where Max_Iter is set to $n \times 500$ if $n \leq 100$; otherwise $n \times 250$.

4.5.3 Parameter tuning

The proposed ILS has 4 parameters ($Para_{pert}$, $Para_{dpr}$, $Para_{pb}$, and $Para_{iter}$) which needs to be tuned. Since tuning of parameters is a tough task. so, we considered potential values of each parameter (mainly for three parameters $Para_{pert}$, $Para_{dpr}$, and $Para_{pb}$ reported in Table 4.2) which are based on our initial study as well as our experience. Such potential values does not cover all values, but help in finding solutions of high quality. During tuning, we change one potential value of a parameter, while keeping other parameters to their final values. Table 4.2 summarizes the results of parameter tuning. Here, results returned by ILS mean the average makespan of each category of *MOD set of instances* over single run. Based on these results, We designate $Para_{pert}$ as 0.3, $Para_{dpr}$ as 0.3, and $Para_{pb}$ as 0.1 as the concluding values for each respective parameter (emphasized in bold font in Table 4.2). The final value of $Para_{iter}$, after our preliminary observation, is set to n if $n \leq 50$; otherwise 500.

Table 4.2: Effect of parameter tuning on the quality of solutions

Parameters	Size														
	Values	**10**	**20**	**30**	**40**	***50***	***60***	***70***	***80***	***90***	***100***	***150***	***200***	***250***	***300***
P_m	0.2	267.48	519.5	789.88	1065.28	**1326.90**	1561.58	1776.74	2074.64	2354.24	2587.78	3854.32	5166.70	**6379.70**	7663.52
	0.3	**267.48**	**519.50**	**789.88**	**1065.26**	1326.94	**1561.56**	**1776.52**	**2074.58**	**2354.16**	**2587.72**	**3854.32**	**5166.70**	6379.78	**7663.46**
	0.4	267.48	519.50	789.88	1065.26	1326.94	1561.68	1776.64	2074.60	2354.22	2587.80	3854.34	5166.72	6379.78	7663.46
Pert	0.2	267.48	519.50	789.88	1065.32	**1326.92**	1561.60	1776.64	2074.66	2354.20	2587.80	3854.32	5166.74	6379.68	7663.50
	0.3	**267.48**	**519.5**	**789.88**	1065.26	1326.94	**1561.56**	**1776.52**	2074.58	**2354.16**	**2587.72**	**3854.32**	**5166.70**	6379.78	7663.46
	0.4	267.48	519.50	789.88	**1065.24**	1326.96	1561.64	1776.52	**2074.56**	2354.18	2587.72	3854.32	5166.72	**6379.64**	**7663.44**
P_s	0.05	267.48	519.50	789.90	1065.26	**1326.92**	1561.90	1776.52	2074.58	2354.22	2587.74	3854.32	5166.74	6379.76	7663.46
	0.10	**267.48**	**519.50**	**789.88**	**1065.26**	1326.94	**1561.56**	**1776.52**	**2074.58**	**2354.16**	**2587.72**	**3854.32**	**5166.70**	6379.78	**7663.46**
	0.15	267.48	519.50	789.88	1065.26	1326.96	1561.56	1776.60	2074.56	2354.20	2587.72	3854.32	5166.82	**6379.70**	7663.52

4.5.4 Comparison of our ILS against state-of-the-art Metaheuristic approaches

We present a summary of the computational results on each category of *MOD* set of benchmark instances and give a general overview of the performance of our proposed ILS. The summary is presented in Table 4.3 which shows detailed comparative results on the *MOD* instances between ILS and other methods including 19 approaches [37] and GA [38]. Results are reported with respect to average relative percentage deviation (ARPD) for each category of instances, where each category is composed of 50 instances. The column $TOTAL$ represents the average of ARPD on all categories of instances obtained by its respective method. The best values $ARPD$ and $TOTAL$ are are emphasized in bold font.

Table 4.3 clearly shows that the performance of ILS is superior to all 19 approaches except three categories of instances with n = 10, 20, 30, where the performance of ILS is equivalent to that of MILP. Even, the $TOTAL$ value of ILS (i.e. -3.723) is superior to that of all 19 approaches. When comparing with GA, ILS performs better, as the $TOTAL$ value of ILS (i.e., -3.723) is better than that of GA (i.e., -3.696). In fact, ILS discovers new best results on 6 out of 14 categories of *MOD* instances.

On computational time part, Table 4.4 reports the running time (in seconds) to reach the best solution by the top 3 methods (NEW_BF, NEW_FF and MILP) [37], GA [38], and ILS on each category of *MOD* instances. The running time of NEW_BF, NEW_FF, and MILP reported in Table 4.4 are extracted from the source mentioned in [37], whereas the running time of GA is extracted from the source mentioned in [38]. The column $Total$ presents the average running time for all categories of *MOD* instances obtained by each respective method. One can infer from the $Total$ column of Table 4.4 that ILS is many times faster than all three methods (NEW_BF, NEW_FF and MILP) and is faster than GA.

It is to be noted that the performance of ILS and GA on *LOW* instances are similar, as both approaches find the same APRD value on each category of *LOW* instances. That is why, we have not discussed about these *LOW* instances.

Table 4.3: ARPD value returned by each approach on each category of MOD instances

Method	Size of each category of MOD instances														TOTAL
	10	**20**	**30**	**40**	***50***	***60***	***70***	***80***	***90***	***100***	***150***	***200***	***250***	***300***	
NEW	4.314	7.775	8.395	17.005	13.464	12.382	13.151	15.351	18.182	14.780	21.734	21.685	24.622	27.919	15.768
NEW BF	0.151	0.422	0.474	0.384	0.365	0.459	0.241	0.172	0.101	0.112	0.031	0.007	0.008	0.003	0.209
NEW FF	0.206	0.448	0.607	0.450	0.510	0.557	0.313	0.223	0.155	0.122	0.043	0.028	0.021	0.017	0.264
BFR	7.841	7.318	7.129	5.977	5.899	5.808	5.239	5.264	4.290	4.054	3.419	3.103	2.933	2.732	5.072
BFD	1.846	1.293	0.871	0.691	0.671	0.655	0.365	0.285	0.237	0.157	0.050	0.036	0.030	0.023	0.515
BFI	21.292	25.256	28.502	29.821	31.210	31.808	30.809	30.312	30.885	30.889	31.335	32.628	31.755	32.235	29.910
BFHILO	6.198	6.802	7.182	7.679	9.439	8.485	7.950	8.010	8.136	8.011	7.052	8.010	8.000	8.854	7.843
BFLOHI	7.314	8.168	7.960	8.287	10.205	9.486	8.446	8.305	8.426	8.449	7.288	8.173	8.285	8.935	8.409
BFV	7.740	11.858	12.958	13.267	13.853	14.943	14.281	14.117	14.501	14.509	14.855	15.947	15.423	15.777	13.859
BFA	8.177	8.280	8.554	8.848	8.054	8.301	6.847	6.631	7.202	5.443	5.762	7.140	7.520	7.208	7.426
FFR	7.812	8.747	8.335	7.114	6.704	6.225	5.637	6.364	5.101	4.725	4.118	3.864	3.653	3.685	5.863
FFD	1.825	1.191	0.875	0.621	0.653	0.658	0.367	0.283	0.221	0.156	0.045	0.037	0.026	0.021	0.499
FFI	21.292	25.256	28.502	29.821	31.210	31.808	30.809	30.312	30.885	30.889	31.335	32.628	31.755	32.235	29.910
FFHILO	6.000	6.930	7.245	7.679	9.397	8.542	8.001	8.025	8.200	8.116	7.052	8.090	8.032	8.879	7.871
FFLOHI	7.449	8.716	8.152	8.287	10.207	9.654	8.559	8.305	8.492	8.527	7.288	8.206	8.302	8.948	8.507
FFV	7.740	12.074	12.958	13.360	13.853	15.018	14.281	14.117	14.620	14.549	14.829	15.929	15.423	15.805	13.897
FFA	7.360	7.708	8.496	8.776	8.111	8.233	6.919	6.668	7.261	5.465	5.787	7.150	7.526	7.227	7.335
MFFD	2.084	1.880	1.643	1.698	1.311	1.723	0.970	1.186	1.219	0.765	0.687	0.694	0.848	0.924	1.259
MILP	**0.000**	**0.000**	**0.000**	0.002	0.000	0.001	0.025	0.010	0.052	0.128	0.344	0.762	0.741	0.850	0.208
GA	**0.000**	**0.000**	**0.000**	-0.003	-0.023	-0.026	-0.088	-0.101	-0.193	**-0.263**	**-0.433**	**-0.856**	**-0.808**	**-0.903**	-3.696
ILS	**0.000**	**0.000**	**0.000**	**-0.008**	**-0.029**	**0.029**	**-0.096**	**-0.104**	**-0.198**	-0.262	**-0.433**	**-0.856**	-0.806	-0.902	**-3.723**

Table 4.4: Running time (in seconds) taken by each approach on each category of MOD instances

Method	Size of each category of MOD instances														TOTAL
	10	**20**	**30**	**40**	***50***	***60***	***70***	***80***	***90***	***100***	***150***	***200***	***250***	***300***	
NEW BF	0.001	0.003	0.007	0.018	0.040	0.069	0.110	0.164	0.265	0.345	1.261	3.784	10.888	23.527	2.892
NEW FF	0.000	0.002	0.007	0.016	0.031	0.059	0.087	0.134	0.222	0.293	1.058	2.902	9.093	19.192	2.364
MILP	0.091	1.256	34.289	97.395	241.888	400.628	471.155	735.963	834.422	811.741	874.694	900.409	900.878	901.630	514.746
GA	0.000	0.010	0.020	0.040	0.050	0.080	0.100	0.130	0.150	0.170	0.370	0.680	1.110	1.680	0.613
ILS	0.000	0.010	0.020	0.030	0.050	0.080	0.110	0.130	0.170	0.220	0.320	0.560	1.010	1.680	**0.314**

4.5.5 Statistical test

In this subsection, statistical test is evaluated on ARPD values (results) returned by our proposed ILS with that of top 3 methods (NEW_BF, NEW_FF, and MILP) [37] and GA [38] for the PMASMSP. We conducted Wilcoxon's Signed-Rank test using a two-tailed nonparametric approach with a significance level of 0.05 on each category of the *MOD* instances. Its results summarized in Table 4.5 demonstrate that ILS statistically performs better than all methods except GA.

Table 4.5: Statistical results

Approaches	Significant
ILS vs NEW_BF	Yes
ILS vs NEW_FF	Yes
ILS vs MILP	Yes
ILS vs GA	No

4.5.6 A study on the use of 2nd step procedure in the acceptance criterion of ILS

The 2nd step procedure in the acceptance criterion of ILS is applied only when the 1st step procedure repeatedly uses the same current solution as the new starting solution over a subsequent number of $Para_{iter}$ iterations. When this occurs, it is presumed that the current solution of ILS is trapped in a deep local optimum, rendering the intensification of ILS ineffective. To address this, there is a need to switch from intensification to diversification, and then resume intensification for ILS. For diversification, the 2nd step procedure in the acceptance criterion is applied.

In this subsection, we analyze the importance of the 2nd step procedure in the acceptance criterion of ILS and examine how the solution quality is affected when this procedure is completely omitted (referred to as ILS-2ndStep). We tested both the main ILS approach and ILS without the 2nd step procedure (ILS-2ndStep) on each category

Table 4.6: Results of ILS and ILS-2ndStep on each category of *MOD* instances

Size of each category of *MOD* instances	ILS	ILS-2ndStep
10	**267.48**	267.74
20	**519.5**	519.80
30	**789.88**	791.24
40	**1065.26**	1066.92
50	**1326.94**	1327.60
60	**1561.56**	1562.52
70	**1776.52**	1779.00
80	**2074.58**	2076.60
90	**2354.16**	2357.74
100	**2587.72**	2588.86
150	**3854.32**	3856.32
200	**5166.70**	5172.72
250	**6379.78**	6384.24
300	**7663.46**	7667.84

of *MOD* instances. The results, shown in Table 4.6, report the average makespan for each category of *MOD* instances over a single run, with the better results highlighted in bold font. From the results in Table 4.6, ILS-2ndStep finds inferior quality solutions compared to the main ILS approach which uses the 2nd step procedure in the acceptance criterion on all categories of *MOD* instances. One can conclude that the 2nd step procedure in the acceptance criterion of ILS plays a significant role in achieving high-quality solutions.

We further investigate the significance of replacing the current solution with an initial solution during the initial solution procedure (see Section 4.4.6) with a probability ($Para_{pb}$) in the second step of the acceptance criterion of ILS. Specifically, we examine how the solution quality is affected when we use the *DRP* procedure (see Section 4.4.6) exclusively in the second step of the acceptance criterion, completely ignoring the replacement of the current solution with an initial solution (referred to as ILS-initialization). We tested ILS-initialization on each category of *MOD* instances. The results, reported in Table 4.7, show the average makespan for each category of *MOD* instances over a single run, with better results emphasized in bold font. From the results in Table 4.7, ILS with a mixed strategy of replacing the current solution through either the initial solution procedure (with a probability $Para_{pb}$) or the *DRP* procedure followed by a repair operator (with a probability 1-$Para_{pb}$) helps in finding high-quality solutions as compared to ILS-initialization on most categories of *MOD* instances.

Table 4.7: Results of ILS and ILS-initialization on each category of *MOD* instances

Size of each category of *MOD* instances	ILS	ILS-initialization
10	**267.48**	267.64
20	**519.50**	519.52
30	**789.88**	**789.88**
40	**1065.26**	**1065.26**
50	1326.94	**1326.90**
60	**1561.56**	1561.62
70	**1776.52**	1776.58
80	**2074.58**	2074.62
90	**2354.16**	2354.20
100	**2587.72**	2587.76
150	**3854.32**	**3854.32**
200	**5166.70**	5166.72
250	6379.78	**6379.76**
300	**7663.46**	**7663.46**

Clearly, the overall efficiency of ILS is quite sensitive to the acceptance criterion applied. This additional diversification mechanism, i.e., the use of the 2nd step procedure in the acceptance criterion of ILS, significantly improves the performance of ILS.

Chapter 5

Min-degree constrained minimum spanning tree problem

5.1 Introduction

In graph, the min-degree constrained minimum spanning tree (MDMST) problem [106, 107] is well-known variant of minimum spanning tree problem that imposes specific degree constraints on the vertices within the spanning tree. In particular, the MDMST problem seeks a spanning tree of a given undirected, connected, and weighted graph $G = (V, E, w)$, where V represents the set of vertices, E denotes the edges, and w indicates the positive weights associated with the edges. The objective is to find a spanning tree where the total cost is minimized, while ensuring that each non-leaf vertex in the spanning tree has a degree of at least d, where d is a given positive integer. This degree constraint differentiates the MDMST problem from other traditional minimum spanning tree problems, as it places additional structural requirements on the spanning tree.

Similar to previous studies of the MDMST problem [39, 41, 42, 44, 108, 109, 110], we also consider the graph G to be complete in our approach.

The MDMST problem shares similarities with the degree-constrained minimum spanning tree problem, which focuses on constructing a spanning tree of minimum cost, where each node adheres to a predefined degree limit [92]. In addition to this, there are several related problems that consider degree-based criteria, such as minimizing the number of branch vertices or minimizing the degree sum of branch nodes [111, 112]. A recent variation of the MDMST problem has also been explored, where specific nodes are predetermined as central or terminal, and these roles are fixed in advance [113].

The MDMST problem belongs to the class of $\mathcal{NP}$-hard problems and has significant relevance across various network design applications. It plays an important role in domains like VLSI circuit design, data clustering, and telecommunications [114, 115, 116], where the MDMST problem seeks to design a minimal-cost network that includes designated nodes with degrees above a certain threshold, enabling the placement of centralized information-handling devices.

5.1.1 Literature Review

The MDMST problem has attracted considerable attention in the research community, leading to the development of both exact algorithms and heuristic strategies. Almeida et al. [106, 107], who first introduced the problem, established its $\mathcal{NP}$-hardness for the case where $\frac{n}{2} > d \geq 3$. They proposed integer programming models based on single- and multi-commodity flow formulations, and applied branch-and-bound methods to solve them. Building on this work, Akgun and Tansel [117] proposed an alternative formulation incorporating Miller–Tucker–Zemlin (MTZ) constraints, also solved using a branch-and-bound framework. Subsequently, Martinez and Cunha [40] developed another variant of the formulation and presented a branch-and-cut approach (denoted by BC) for this problem.

Besides the development of exact algorithms for solving the MDMST problem, researchers have increasingly turned to metaheuristic approaches due to the scalability issues associated with exact methods. As the size of the problem instances grows, exact techniques often become computationally impractical. Metaheuristics, while not guaranteeing optimal solutions, have demonstrated the ability to deliver high-quality results within reasonable time frames.

Among these approaches, four VNS variants were proposed by Martins and Souza [39], namely: VNS/ESO, Rk-VNS/ESO, Rk-VNS/ESO1, and Rk-VNS/greedy. In the first variant, an enhanced version of a second order repetitive technique (ESO) [118] is integrated into VNS framework, replacing the most common local search operator in the improvement phase. In the second and third variants, random versions of the first variant are applied, while the fourth variant incorporates a greedy heuristic guided by Kruskal's algorithm. All four variants employ a neighborhood structure denoted by the sequence $N^{k_{min}}, \ldots, N^{k_{max}}$, with $k_{min} \geq 1$, $k_{max} \leq |V|$-1 and $k_{min} < k_{max}$. In each VNS, the iteration begins with a current solution T and a value k within this range. During the shaking phase, k edges are randomly removed from T and a new tree is generated using Kruskal's heuristic. This is followed by two applications of the ESO for further improvement in its solution quality.

Later, two metaheuristic techniques – artificial bee colony algorithm and ant colony optimization approaches – combined with local search strategies were proposed [108, 109]. Salgueiro et al. [41] investigated generational genetic algorithms (referred to as min-GA) and proposed three tailored versions (*gen-0*, *gen-1*, and *gen-2*) for the MDMST problem. These variants primarily differed in their genetic representations and the specific application of genetic operators (crossover and mutation operators).

Singh and Sundar [42] proposed a hybrid steady-state genetic algorithm (HSSGA), which integrated genetic operators with a fast local search. Their experimental results indicated that HSSGA was superior as compared to existing methods. Later, Ghoshal and Sundar [110] presented two metaheuristic approaches for the MDMST problem, in which the first one is ABC algorithm that combines with local search strategies and iterated local search. These approaches demonstrated some improved results against existing approaches. Recently, Wu *et al.* [44] introduced a population-based simulated annealing (SA) approach in which three neighborhood operators – (i) construction-based neighbor operator, (ii) repaired-based neighbor operator, and (iii) learning-based neighbor operator – are used. To further improve its solution quality, a fast local search was applied. It is to be noted that the second neighborhood operator – repaired-based neighbor operator – and fast local search operator are taken from [42]. Results obtained by PSA show that it is superior to existing approaches on larger instances at the cost of high computational time.

5.1.2 Our Contribution

The MDMST problem poses significant computational challenges, particularly as the size of the instance grows. Drawing insights from the broader literature on combinatorial optimization problems (COPs), variable neighborhood search (VNS) approach has demonstrated considerable success across a range of $\mathcal{NP}$-hard problems [31, 32, 33]. This suggests that VNS could serve as an effective and promising strategy for tackling the MDMST problem as well. One of the key strengths of VNS approach lies in its flexible and powerful framework, which systematically explores multiple neighborhood structures to escape local optima and guide the search toward improved solutions. However, the effectiveness of VNS critically depends on the design and selection of neighborhood structures. While many types of neighborhoods can be defined, inappropriate or overly generic choices may not lead to an effective search, thereby limiting the potential for meaningful improvement.

Motivated by this observation, we propose an improved version of VNS framework (referred to VNS^I) for solving the MDMST problem. In contrast to the existing four VNS variants [39] for the MDMST problem, which rely on random edge removal across a sequence of neighborhoods ($N^{k_{min}}, \ldots, N^{k_{max}}$, with $k_{min} \geq 1$, $k_{max} \leq |V|$-1 and $k_{min} < k_{max}$) and use Kruskal's algorithm followed by two applications of local search for solution improvement, our approach introduces structured and guided neighborhood designs. Specifically, our proposed VNS^I defines four distinct neighborhoods $N_1, \ldots, N_4$, each tailored to remove edges based on the degrees of their endpoints. This degree-informed criterion enables VNS^I to focus on edges with greater impact on node degrees.

The proposed neighborhood structures are arranged to progressively the search – from broader exploration to more narrow one – allowing for a balanced traversal of the solution space. When combined with local search strategies, this approach enhances both diversification and exploitation, fostering deeper solution refinement. Experimental results show that the proposed VNS^I is superior to the existing four VNS variants, and when compared to other state-of-the-art approaches, VNS^I is competitive with some approaches on CRD and SYN data-sets (instances of smaller size); however, VNS^I is superior to all approaches on ALM and E250 data-sets (instances of larger size). Notably, it consistently finds better solutions, including several new best-known results, especially on larger problem instances. This underscores the value of incorporating problem-specific knowledge into neighborhood structure design and highlights the effectiveness of our approach in addressing the MDMST problem.

The structure of this chapter is as follows: In section 2, we present our proposed VNS^I for the MDMST problem. Section 3 provides a comprehensive evaluation of VNS^I against state-of-the-art methods.

5.2 Variable neighborhood search for the MDMST problem

We have adopted a general variable neighborhood search framework (VNS^I) for the MDMST problem. In our proposed VNS^I framework, we define a sequence of neighborhood operators N_k for $k = 1, \ldots, K_{max}$ (where $K_{max} = 4$), each representing a specific neighborhood structure and collectively determining the order of neighborhood exploration. This combination effectively balances diversification and intensification, enabling a deeper and systematic exploration of the solution space to identify improved solutions. Starting with an initial solution (say T), a certain number of cycles (also a stopping criterion) is executed. In each cycle, the procedure systematically explores the neighborhood structures associated with the operators N_k for $k = 1, \ldots, k_{max}$, in the pre-defined sequence and followed by local search strategies, leading to a local optimum T''. If T'' is superior to the current solution T, T'' takes the place of T, and the procedure recenters its search from the first neighborhood N_1.

When the best-so-far solution does not improve over a fixed number of iterations, VNS^I performs a reset in which the current solution is replaced by a newly generated solution. This strategy of restarting the search from a different point in the solution space increases the chances of discovering higher-quality solutions.

The proposed VNS^I framework tailored to the MDMST problem is presented in the Algorithm 13.

Algorithm 13: The pseudo-code of VNS^I framework tailored to the MDMST problem

```
Initialize with generating an initial solution T;
Best ← T;
while Termination criterion is not fulfilled do
    for k ← 1 to K_max do
        // Generate a neighborhood solution T' in the k-th
        neighborhood structure N_k of the current solution T;
        T' ← Neighborhood_Operator(T, N_k);
        T'' ← LocalSearch_Strategies(T');
        // Acceptance Criterion;
        if F(T) > F(T'') then
            T ← T'';
            k ← 1;
            if T is better than Best then
                Best ← T;
        else
            k ← k + 1;
    Apply Re-start Criterion;
```

The proposed VNS^I framework for the MDMST problem is structured into four main components: (a) initial solution generation, (b) neighborhood structures, (c) local search improvement strategy, (d) acceptance criterion, and (e) re-start criterion, as described in Algorithm 13.

5.2.1 Solution Representation

The solution to the MDMST problem is represented as a spanning tree of the graph $G(V, E)$, where each vertex has a degree constraint. We use an edge-set encoding [92] to represent the solution, which consists of a set of edges that form a spanning tree. Each solution is encoded as a set of $|V|$-1 edges, where each element corresponds to an edge between two vertices. This encoding supports high flexibility and locality with the problem-specific neighborhood operators.

5.2.2 Initial Solution Generation

The procedure for constructing an initial solution uses a modified variant of Prim's algorithm [93], subject to a minimum degree constraint. Rather than consistently selecting an edge with the minimum cost from the set of candidate edges, the procedure employs a hybrid strategy that balances greed and randomness. At each step, it either picks the edge with the lowest cost or selects one at random from the candidate list with equal probability. An edge is called a candidate edge if it connects a vertex v_1 already included in the partial tree T to a vertex v_2 not yet included, provided that: (i) v_1 has at least d incident edges within T, and (ii) there are enough remaining adjacent unconnected vertices to v_1 to fulfill its minimum degree requirement. This mixed selection strategy promotes solution diversity while ensuring that the search initializes with a promising initial solution.

It is to be noted that such procedure for constructing an initial solution was also applied in [110].

5.2.3 Neighborhood structures

The design of diverse neighborhood structures in VNS plays a crucial role in adapting the search behavior and helps in identifying solutions of high quality in the search space.

In the literature, neighborhood operators for the MDMST problem are often designed by randomly removing one or more edges from the current solution, followed by the application of repair strategies to restore feasibility, followed by local search strategies. The specific repair mechanisms vary across studies [39, 41, 42]. In contrast, our approach adopts a more systematic and guided methodology: each of our neighborhood structures selects a specific edge for removal based on the degree of its endpoints, thus introducing a problem-aware perspective to neighborhood design. Building on this principle, our proposed VNS framework for the MDMST problem integrates four predefined and structurally distinct neighborhood structures, denoted as $N_1, \ldots, N_k$, each designed to guide the search from broader to narrower regions of the solution space by progressively focusing on edges with more localized impact, especially when coupled with local search strategies. Here K is 4. This combination effectively balances diversification and intensification, enabling a deeper and systematic exploration of the solution space to identify improved solutions. The description of each neighborhood structure is as follows:

- □ In the first neighborhood structure (N_1) of VNS framework, the operator randomly selects a specific edge (say e_{pq}) randomly from the current solution T that satisfies the condition $((deg[p] == d \&\& deg[q] \geq d) \,||\, (deg[p] \geq d \&\& deg[q] == d))$. Assume that the degree of vertices p and q of selected edge e_{pq} are d and greater than d, respectively. The selected edge e_{pq} is then removed from T, thereby partitioning the solution into two disconnected sub-trees, denoted T_1 and T_2. To reconnect these components, a new edge (e_{xy}), distinct from e_{pq}, is randomly selected from the set $G \setminus E_T$, such that vertex $x \in T_1$ and vertex $y \in T_2$. Although this step restores the connectivity of the tree, it may introduce two types of constraint violations: (i) vertex x might not yet meet the required degree d, and (ii) the degree of vertex p may fall below d due to the deletion of e_{pq}, thereby leaving the solution infeasible. If the first constraint violation occurs, a set of vertices adjacent to p – all except one – are randomly selected and reconnected to vertex x, making the solution feasible. This transformation turns p into a leaf vertex and satisfies the minimum degree constraint for x. However, if the first constraint violation does not hold, there are two simple strategies to satisfy the minimum degree constraint of p. In the first strategy, if there exist at least two vertices with degree greater than d, a fixed number of their adjacent leaf vertices are removed and reconnected to p, forming new leaf edges that fulfill its degree constraint. In the second strategy, similar to the previous case, all but one of the vertices adjacent to p are randomly reassigned to x, making p a leaf node and thereby making the solution feasible. An analogous procedure is applied if the degrees of vertices p and q of selected edge e_{pq} are greater than d and d, respectively.

- □ In N_2, the operator randomly selects an edge (say e_{pq}) from the current solution T that satisfies the condition $(deg[p] > d \&\& deg[q] > d)$. The selected edge e_{pq} is then removed from T, thereby partitioning the solution into two disconnected sub-trees, denoted T_1 and T_2. To reestablish connectivity, a new edge (e_{xy}), distinct from e_{pq}, is randomly selected from the set $G \setminus E_T$, such that vertex $x \in T_1$ and vertex $y \in T_2$. However, this reconnection may still leave the solution infeasible if either x or y does not meet the minimum degree requirement d. If vertex x

violates the constraint, the operator searches for leaf vertices connected to other vertices in T whose degrees exceed d. These leaf vertices are then disconnected and reassigned to x as its leaf vertices, incrementally increasing its degree. If this adjustment is insufficient to satisfy the degree requirement, then all but one vertices adjacent to p are randomly redirected to x, effectively transforming p into a leaf vertex and restoring feasibility. An analogous procedure is applied if the degree constraint violation is observed at vertex y.

- In N_3, the operator randomly selects an edge (say e_{pq}) from the current solution T that satisfies the condition $((deg[p] == 1 \&\& deg[q] == d) \| (deg[p] == d \&\& deg[q] == 1))$. Assume that the degree of vertices p and q of selected edge e_{pq} are 1 and d, respectively. The selected edge e_{pq} is removed from T, thereby partitioning T into two disconnected sub-trees, T_1 and T_2. Importantly, T_1 initially contains only the vertex p. Before reconnecting these sub-trees, the operator checks the number of vertices with a degree $\geq d$ in T_2. If such vertices exist, the sub-trees are reconnected with probability (P_e) by selecting a minimum-cost edge e_{py} from among all candidate edges that connect vertex p in T_1 to a vertex y in T_2 with a degree $\geq d$. Otherwise, an edge e_{py} is selected randomly from the candidate edges with probability (1-P_e), where P_e is a parameter. To address the degree of vertex q, which falls below d, we employ two strategies. First, the operator attempts to locate a leaf vertex (not p) connected to y. If such a vertex is found, it is disconnected from y and connected to q, thus restoring feasibility. If such a vertex is not found, the operator redirects all but one of the vertices adjacent to q to y, effectively transforming q into a leaf vertex and restoring feasibility.

  If there are no vertices in T_2 with a degree $\geq d$, the sub-trees are reconnected with probability (P_e) by selecting a minimum-cost edge e_{px} from among all candidate edges that connect vertex p in T_1 to a vertex x adjacent to q in T_2. Otherwise, an edge e_{px} is selected randomly from the candidate edges with probability (1-P_e), where P_e is a parameter.

- In N_4, the operator randomly selects an edge (say e_{pq}) from the current solution that satisfies the condition $((deg[p] == 1 \&\& deg[q] > d) \| (deg[p] > d \&\& deg[q] == 1))$. The selected edge e_{pq} is then removed from T, thereby partitioning the solution into two disconnected sub-trees, denoted T_1 and T_2.

Importantly, T_1 initially contains only the vertex p. Before reconnecting these sub-trees, the operator checks the number of vertices with a degree $\geq d$ in T_2. If such vertices exist, the sub-trees are reconnected with probability (P_e) by selecting a minimum-cost edge e_{py} from among all candidate edges that connect vertex p in T_1 to a vertex y in T_2 with a degree $\geq d$. Otherwise, an edge e_{py} is selected randomly from the candidate edges with probability (1-P_e), where P_e is a parameter.

If there are no vertices in T_2 with a degree $\geq d$, the sub-trees are reconnected with probability (P_e) by selecting a minimum-cost edge e_{px} from among all candidate edges that connect vertex p in T_1 to a vertex x adjacent to q in T_2. Otherwise, an edge e_{px} is selected randomly from the candidate edges with probability (1-P_e), where P_e is a parameter.

It is important to note that the procedure rarely fails to find an edge associated with its neighborhood structure. However, if this does occur – even after scanning all edges in the solution – the procedure proceeds to the next neighborhood structure in order to identify a suitable edge.

5.2.4 Local search strategies

We employ two local search strategies, referred to as LS1 and LS2, applied in a cyclic sequence: LS1 $\rightarrow$ LS2 $\rightarrow$ LS1 $\rightarrow$ LS2, ..., LS1 $\rightarrow$ LS2. This cyclic application is designed to intensify the search in the vicinity of a solution generated via a neighborhood structure (outlined in Section 5.2.3). This cycle continues until LS2 fails to produce any further improvements in the solution. The solution obtained after applying the local search procedures to T' is denoted as T''. The use of LS1 and LS2 in our methodology builds on previous work conducted in collaboration with a former student, where these strategies were initially introduced in [110] and [42]. The concept of applying LS1 and LS2 cyclically is also taken from [110].

The first local search method, LS1, also known as Fast Local Search (FLS), focuses on specific types of swap and insert operations applied to T'. These operations are designed to always yield feasible solutions, thus eliminating the need for repair mechanisms and enhancing computational efficiency. The swap operation selects two leaf-edges from T' and attempts to replace them with two other edges from $E \setminus T'$ (excluding those in T') that also qualify as leaf-edges in T', with the goal of improving the solution. The insert operation, which includes five specific cases, is based on inserting a new edge while removing an internal edge from T', aiming to minimize the overall cost of the resulting tree. For more detailed information on FLS, readers can refer to [42]. Alternatively, instead of executing swap and insert operations until no improvement (as done in [42]), each operation is limited to $|V|$-1 iterations, where in each iteration, an edge from T' is chosen in order for a potential swap or insert move. It is to be noted that FLS was also coupled with the neighborhood operators for further solution improvement in simulated annealing framework [44].

The second local search, LS2, is based on a two-edge replacement mechanism, as introduced in [119]. In earlier work, LS2 was executed for $\frac{|V|}{2}$ iterations until no improvement could be achievable. In contrast, following the approach of [110], our proposed VNS^I applies LS2 for $|V|$-1 iterations. During each iteration, a randomly selected edge e_{ij} from T' is chosen. Among all feasible candidate edges that are non-adjacent to e_{ij}, LS2 identifies an edge e_{kl} such that replacing both e_{ij} and e_{kl} with e_{ik} and e_{jl}, where e_{ik}, $e_{jl} \in E \setminus T'$, results in the maximum possible improvement in the current solution T'.

5.2.5 Re-start Criterion

If the best-found solution remains unchanged for a predefined number of iterations (denoted as P_{ReS}), our proposed VNS^1 performs a reset by generating a new solution using the N_1 neighborhood structure (refer to Section 5.2.3 for further details). This strategy of restarting the search from a different point in the solution space increases the chances of discovering higher-quality solutions. While we also tested alternative methods – including neighborhood structures N_2, N_3, N_4, and the initial solution construction technique – for generating a new starting solution, our preliminary results indicate that N_1 is more effective in yielding better-quality solutions. This may be attributed to N_1's ability to introduce significant changes to the current solution (i.e., strong perturbation), allowing the search to explore new regions without overly disrupting the underlying structure of the solution.

The interplay between intensification (searching deeper in promising areas) and diversification (exploring new areas) is a crucial aspect of VNS. This balance ensures a thorough exploration of the solution space while also exploiting promising regions for better solutions.

5.3 Computational Results

In this section, we assess the performance of our proposed VNS^I for the MDMST problem by comparing it against state-of-the-art approaches using benchmark data-sets.

5.3.1 Benchmark Instances

Our proposed VNS^I is tested on four distinct benchmark data-sets consisting of 105 instances for the MDMST problem. These four data-sets are CRD, SYM, ALM, and E250. Table 5.1 presents the description of instances of each data-set. In Table 5.1, first column (denoted by Data-set) is the name of the data-set; second column (denoted by No_inst) is the number of instances; third column (denoted by Vertex Range) is the range of vertices of instances associated with its data-set; and fourth column (denoted by Description) is the description of how the instance is generated.

5.3.2 Experimental setup

The proposed VNS^I is implemented in C and executed on a Linux-based PC featuring a 3.2 GHz quad-core Intel Core i5 processor and 4 GB of RAM to conduct computational experiments on benchmark data-sets for the MDMST problem. As a termination criterion, VNS^I ceases execution when no improvement in the best solution is observed for Max_iter consecutive iterations, where $Max\_iter = d \times 1000$ if $|V| \leq 50$; $Max\_iter = d \times 500$ if $70 \leq |V| \leq 100$; and $Max\_iter = d \times 250$ if $|V| > 100$.

5.3.3 Parameter tuning

To ensure VNS^I delivers its best performance, careful parameter tuning is critical. VNS^I involves three parameters: K_{max}, P_init, and P_{ReS}. Since we explore all neighborhood structures associated with the operators N_k for $k = 1, \ldots, K_{\max}$ in order to delete an edge, we set $K_{\max} = 4$. This leaves two parameters, P_e and P_{ReS}, to be tuned. For this, we consider the ranges $P_e = \{0.2, 0.3, 0.4, 0.5, 0.6, 0.7\}$, and $P_{ReS} =$

Table 5.1: Details of each data-set

Data-set	No_inst	Vertex range	Description
CRD	30	{30,50,70,100}	Vertices are randomly distributed inside a square, with edge weights representing the Euclidean distances between connected vertex pairs.
SYM	24	{30,50,70}	Similar to the CRD instances, these points are generated in a high-dimensional Euclidean space.
E250	15	250	These are Euclidean graphs where vertices are uniformly distributed within a unit square, and each edge is weighted by the Euclidean distance between its two endpoints.
ALM	36	{100,200,300,400,500}	Vertices are uniformly distributed over a 480 × 640 grid, with edge weights computed as the truncated Euclidean distance between their endpoints.

$\{100, 200, 300, 400\}$. By evaluating all parameter combinations across a set of different instances, we identify the best-performing configuration: $P_e = 0.4$ and $P_{ReS} = 200$.

5.3.4 Comparison of VNS with existing approaches

This subsection provides a detailed empirical evaluation of the proposed VNS^I in comparison with several established approaches.

The existing four VNS variants [39] (see Section 5.1.2 in detail) were evaluated on a Pentium IV 3.2 GHz processor with 512 MB RAM and executed for a single run per benchmark instance. In a separate study, authors [41] tested three variants of the min-GA algorithm (*gen-0*, *gen-1*, and *gen-2*) on an Intel Q9550 processor (Core2 Quad 2.83 GHz) with 4 GB of RAM, performing 64 runs per instance. However, complete comparative results across these min-GA variants were not reported in their paper.

Due to the lack of comprehensive result disclosure, it is unclear which of the four VNS variants or the three min-GA variants performs best overall. To resolve this ambiguity, following the approach of [42], we report only the best-performing result among the four VNS variants under the label VNS^O. Similarly, we report the best-performing result among the three min-GA variants – evaluated over 3×64 runs – under the label min-GA*. For the min-GA variants, execution times are presented as ranges (minimum-maximum) across the ALM, SYM, and CRD data-sets.

For BC [40], computational experiments were conducted on a computer equipped with an Intel Core 2 Duo 3 GHz processor and 4 GB of RAM, operating under Linux. The time limits for execution varied based on the instance set: 10,800 seconds for SYM and CRD instances, and 21,600 seconds for ALM instances.

The implementations of HSSGA [42], ILS and hABC [110] were developed in C and tested on a Linux-based machine configured with an Intel Core i5 processor (3.2 GHz, 4 cores) and 4 GB of memory. In contrast, PSA [44] was developed in Java and executed on a Window 11 platform, using a more powerful system featuring an Intel Core i7 processor (3.6 GHz, 8 cores) and 32 GB of RAM. Each of these four approaches – HSSGA, hABC, ILS, and PSA – was independently run 30 times on every benchmark instance. Similarly, our proposed VNS^I approach is also executed 30 times on each benchmark instance.

The best results obtained from the four existing VNS variants and the three min-GA variants have been sourced from [39] and [41], respectively. The results for HSSGA, hABC, and ILS were taken from [42] and [110], respectively, while PSA results were sourced from [44].

Tables 5.2 to 5.3 present the summarized results for the CRD, SYM, and ALM instances, covering all the aforementioned approaches. The columns in these tables

Table 5.2: Results of VNS^I against state-of-the-art approaches on CRD and SYN data-sets

Instance	|V|	d	VNS^O	BC		min-GA*		HSSGA			hABC			ILS			PSA			VNS^I		
			Best	Best	ATET	Best	ATET	Best	Avg	ATET	Best	Avg	ATET	Best	Avg	ATET	Best	Avg	ATET	Best	Avg	ATET
CRD30-1	30	3	4026	4026	0.3	4026	(7.3–19.6)	4026	4026	2.69	4026	4027.17	2.45	4026	4038.63	1.43	4026	4026	2.93	4026	4026	3.38
CRD30-2	30	3	3793	3793	0.2	3793	(7.3–19.4)	3793	3830.97	2.92	3793	3804.3	2.73	3793	3806.87	1.47	3793	3814.77	3.63	3793	3793	6
CRD30-3	30	3	4293	4293	0.2	4293	(7.2–19.3)	4293	4293.93	2.83	4293	4294.13	2.64	4293	4321.77	1.45	4293	4293	3.13	4293	4307.67	0.61
CRD30-1	30	5	5026	5026	0.9	5026	(6.7–20.5)	5026	5028.4	1.93	5026	5026.27	1.85	5026	5035.2	1.19	5026	5026	2.61	5026	5029.23	0.56
CRD30-2	30	5	4648	4648	0.3	4648	(7.2–20.0)	4648	4648	2.41	4648	4648	1.73	4648	4648	1.22	4648	4648	2.55	4648	4742.07	0.69
CRD30-3	30	5	5425	5425	3.2	5425	(7.2–20.5)	5425	5449.13	2.02	5425	5447.5	2.12	5425	5537.53	1.24	5425	5427.8	3.01	5425	5437.27	0.98
CRD50-1	50	3	5512	5512	5	5512	(16.5–43.0)	5512	5539	5.76	5512	5551.7	5.77	5512	5549.7	3.95	5512	5519.6	6.63	5512	5514.07	12.62
CRD50-2	50	3	5813	5813	61.6	5813	(16.4–47.0)	5813	5865.33	5.82	5813	5850.37	5.87	5823	5851.03	3.94	5813	5821.13	10.09	5813	5834	3.42
CRD50-3	50	3	5590	5590	4.2	5590	(–)	5590	5614.4	5.24	5590	5606.43	5.11	5590	5648.87	3.8	5590	5606.53	6.63	5594	5636.23	4.34
CRD50-1	50	5	6908	6908	69	6908	(16.5–44.8)	6908	6928.53	4.93	6908	6916.53	4.57	6908	6958.37	3.4	6908	6919.23	4.69	6908	6935.23	6.31
CRD50-2	50	5	7238	7204	81.2	7244	(16.3–42.7)	7204	7246	4.81	7204	7247.4	4.71	7254	7288.23	3.4	7204	7252.63	7.53	7204	7275.03	8.13
CRD50-3	50	5	7277	7277	589.1	7295	(–)	7277	7349.83	4.39	7277	7308.27	5.24	7283	7358	3.4	7277	7283.4	8.19	7279	7339.13	2.45
CRD50-1	50	10	9633	9633	13.8	9633	(12.7–45.4)	9633	9659.87	3.61	9633	9633.77	3.3	9633	9655.17	2.62	9633	9633	8.78	9633	9633	12.7
CRD50-2	50	10	9743	9743	7.7	9743	(16.6–45.4)	9743	9768.3	3.77	9743	9755.53	4.02	9743	9746.3	2.72	9743	9744.1	10.92	9743	9743	19.06
CRD50-3	50	10	9855	9855	9.1	9855	(–)	9855	9866.03	3.99	9855	9856.73	3.48	9855	9855	2.78	9855	9855	9.95	9855	9857	9.87
CRD70-1	70	3	6609	6516	148.6	6516	(32.5–77.3)	6516	6573.3	9.27	6530	6539.77	9.74	6516	6568.3	7.81	6516	6539.83	14.66	6516	6605.87	10.52
CRD70-2	70	3	6621	6586	4785	6619	(23.1–75.0)	6595	6636.17	10.13	6586	6627.03	11.33	6616	6663.07	7.75	6586	6609.57	13.14	6587	6651.53	12.09
CRD70-3	70	3	7058	7053	3859.9	7053	(25.6–75.7)	7053	7121.4	9.44	7066	7130.7	10.96	7101	7145.37	7.81	7058	7090.57	14.00	7068	7129.6	9.33
CRD70-1	70	5	8177	8144	10800	8139	(26.9–78.1)	8134	8189.6	7.61	8134	8171.8	7.86	8139	8237.43	6.69	8134	8172.1	9.23	8134	8184.5	18.27
CRD70-2	70	5	7971	7943	2416.9	7943	(23.8–74.9)	7943	8104.77	6.85	7943	8029	7.72	7995	8066.4	6.45	7943	7961.53	11.29	7943	8069.27	6.95
CRD70-3	70	5	8628	8419	1402.8	8419	(21.2–77.1)	8421	8540.7	6.92	8421	8502	7.17	8505	8607.43	6.42	8419	8439.07	8.35	8419	8475.63	19.67
CRD70-1	70	10	11355	11235	138.6	11235	(20.7–78.5)	11235	11300	6.56	11235	11300.1	6.55	11235	11312.27	5.99	11235	11272.8	16.95	11235	11245.77	16.81
CRD70-2	70	10	11395	11373	571.9	11373	(28.3–78.2)	11373	11414.6	6.7	11388	11400.1	7.13	11373	11419.13	5.56	11373	11388.3	16.72	11373	11383.07	15.8
CRD70-3	70	10	11986	11979	17.9	11979	(24.7–76.6)	11979	11996.3	6.4	11979	11997.83	6.42	11979	11983.93	5.33	11979	11979	14.70	11979	11979	14.82
CRD100-1	100	5	9387	9352	10 800	9283	(47.2–140.6)	9320	9446.3	13.24	9291	9365.53	15.22	9319	9401.37	13.92	9262	9354.5	15.11	9251	9313.87	20.38
CRD100-2	100	5	9587	9623	10800	9587	(61.2–146.9)	9557	9723.4	14.47	9569	9681.93	14.73	9562	9674.7	13.69	9583	9694.33	18.31	9562	9632.57	23.89
CRD100-3	100	5	9653	9656	10800	9653	(60.5–144.4)	9653	9776.03	13.84	9636	9694.53	13.9	9676	9762.13	13.58	9640	9703.13	15.25	9634	9689.23	21.19
CRD100-1	100	10	13006	12916	5741.6	12916	(51.9–131.5)	12916	13038.3	12.78	12916	13002.1	14.86	12916	12931.33	12.4	12916	12923.57	33.01	12916	12918.1	39.71
CRD100-2	100	10	13255	13032	669.4	13032	(49.7–133.9)	13034	13159.4	12.89	13032	13123.87	13.83	13026	13124.47	11.61	13026	13067.77	31.72	13026	13031.5	34.91
CRD100-3	100	10	13365	13365	1359.4	13365	(42.1–134.9)	13365	13436.1	13.44	13365	13408.03	12.84	13365	13405.9	11.77	13365	13371.33	24.98	13365	13385.5	58.55
SYM30-1	30	3	1197	1197	0.2	1197	(4.7–19.2)	1197	1197.8	2.56	1197	1197	2.47	1197	1212.67	1.48	1197	1204.93	2.60	1197	1222.33	2.05
SYM30-2	30	3	1435	1435	0.1	1435	(4.9–19.1)	1435	1435	2.61	1435	1435	2.69	1435	1444.2	1.53	1435	1444.3	2.51	1435	1472.23	1.85
SYM30-3	30	3	1408	1408	0	1408	(4.6–18.8)	1408	1408	2.7	1408	1408	2.69	1408	1414.43	1.51	1408	1408	2.33	1412	1444.07	1.94
SYM30-1	30	5	1765	1765	0.5	1765	(5.5–19.9)	1765	1765	2.04	1765	1779.93	1.84	1828	1833.97	1.36	1765	1860.53	2.20	1765	1914.87	0.87
SYM30-2	30	5	2092	2092	0.3	2092	(7.2–20.0)	2090	2126.37	2.11	2090	2134.37	1.88	2090	2168.83	1.29	2090	2152.27	2.18	2152	2256.6	0.63
SYM30-3	30	5	2008	2008	0.1	2008	(7.5–19.5)	2008	2008	2.06	2008	2016.4	1.73	2008	2216.6	1.31	2008	2074.43	2.12	2088	2299.03	1.14
SYM50-1	50	3	1278	1278	0.7	1278	(14.9–42.8)	1278	1307.6	5.33	1278	1290.53	5.46	1278	1330.63	3.98	1278	1313.47	4.69	1278	1376.17	5.67
SYM50-2	50	3	1178	1178	0.7	1178	(12.5–43.7)	1178	1201.47	5.5	1178	1188.63	5.69	1188	1230.17	4.04	1187	1207.2	4.85	1207	1291.77	5.81
SYM50-3	50	3	1615	1615	0.2	1615	(13.6–42.9)	1615	1636.77	5.24	1615	1615.97	5.36	1623	1662.97	3.89	1615	1633.17	4.89	1620	1678	6.41
SYM50-1	50	5	2054	2054	5.7	2054	(13.7–40.7)	2071	2199.83	4.17	2114	2142.3	4.09	2114	2171.23	3.51	2054	2215.83	4.24	2108	2304.83	4.14
SYM50-2	50	5	1760	1760	1.3	1760	(8.1–40.8)	1782	1849.6	4.09	1781	1831.1	4	1782	1960.3	3.54	1773	1895.53	4.05	1782	2003.73	4.88
SYM50-3	50	5	2525	2525	10.9	2525	(12.9–43.8)	2541	2592.63	4.08	2538	2558.5	3.95	2564	2650.67	3.58	2541	2611.5	4.15	2572	2721.83	6.64
SYM50-1	50	10	4121	4121	1.6	4121	(–)	4121	4256.33	3.49	4121	4134.97	3.12	4121	4176.2	2.94	4121	4177.93	3.74	4121	4264.23	5.21
SYM50-2	50	10	4166	4166	2.7	4166	(–)	4166	4226.17	3.08	4166	4166	3.01	4166	4187	2.96	4166	4236.7	3.70	4166	4372.67	6.6
SYM50-3	50	10	4979	4979	5.3	4979	(–)	4979	5192.07	3.24	4979	4983.2	3.2	4979	4995.97	3.05	4979	5156.73	4.40	4979	5060.43	6.45
SYM70-1	70	3	1362	1360	1.9	1360	(30.7–70.1)	1360	1390.93	9.22	1364	1379.27	9.83	1369	1405.53	7.52	1362	1385.67	7.56	1443	1588.77	7.71
SYM70-2	70	3	1471	1448	6	1465	(18.9–70.9)	1458	1508.77	8.91	1450	1479.27	9.59	1478	1529.93	7.58	1452	1496.93	7.95	1492	1615.4	7.65
SYM70-3	70	3	1551	1521	4.8	1534	(19.5–72.1)	1521	1554.17	9.15	1521	1526.97	9.56	1545	1582.93	7.76	1521	1554.03	8.78	1558	1661.97	9.19
SYM70-1	70	5	2240	2028	7.3	2087	(18.0–73.4)	2095	2310.73	6.27	2122	2215.67	6.88	2090	2239.67	6.67	2099	2287.27	6.45	2149	2467.6	7.54
SYM70-2	70	5	2496	2165	5	2293	(17.2–73.4)	2187	2467.33	6.55	2229	2349.8	7.1	2254	2427.47	6.59	2183	2484	6.82	2299	2607.93	6.47
SYM70-3	70	5	2242	2210	31.6	2210	(21.8–73.1)	2350	2513.13	6.41	2297	2408.07	6.73	2262	2446.5	6.54	2262	2513.77	6.57	2290	2620.37	6.97
SYM70-1	70	10	5055	4979	115.9	4979	(21.2–76.6)	5055	5464.87	5.73	4991	5173.83	6.15	4979	5198.5	5.69	5055	5333.47	6.30	5055	5241.97	11.1
SYM70-2	70	10	4912	4787	53	4887	(22.6–75.7)	4885	5188.6	5.58	4885	4975.57	5.89	4787	4943.93	5.68	4885	5124.13	7.01	4787	4983.47	9.53
SYM70-3	70	10	5089	4997	112.9	4997	(21.5–75.6)	5141	5461.33	5.52	4997	5189.93	6.12	5098	5236.37	5.62	4997	5373.53	6.15	4997	5296.47	9.89

Table 5.3: Results of VNS^I against state-of-the-art approaches on ALM data-set

Instance	|V|	d	VNS^O	BC		min-GA*		HSSGA			hABC			ILS			PSA			VNS^I		
			Best	*Best*	*ATET*	*Best*	*ATET*	*Best*	*Avg*	*ATET*	*Best*	*Avg*	*ATET*	*Best*	*Avg*	*ATET*	*Best*	*Avg*	*ATET*	*Best*	*Avg*	*ATET*
ALM100-1	100	5	5439	5363	21600	5353	(51.2–129.5)	5318	5403.87	15.57	5323	5371.4	13.89	5318	5376.07	13.58	5330	**5367.07**	17.78	**5317**	5372.2	24.07
ALM100-2	100	5	5207	5072	21600	5022	(59.1–130.8)	5076	5175.9	13.14	5053	5137.07	14.29	5074	5129.9	13.54	**5022**	5056.97	22.41	**5022**	**5053.1**	30.51
ALM100-3	100	5	5456	5457	21600	5441	(55.8–131.8)	5458	5505.73	13.79	5450	5492.77	14.71	5479	5527.77	13.52	**5428**	5469.43	30.92	5433	**5461.17**	33.57
ALM100-1	100	10	7180	**7164**	385.4	7164	(56.3–127.6)	**7164**	7215.7	13.36	**7164**	7207.63	12.49	**7164**	7189.13	11.96	**7164**	7187.33	32.65	**7164**	**7165.87**	35.4
ALM100-2	100	10	6915	**6886**	7706.1	6886	(51.9–128.4)	**6886**	6920.1	13.6	**6886**	6920.67	13.53	**6886**	6902.57	11.83	**6886**	6888.93	34.87	**6886**	**6886.07**	103.89
ALM100-3	100	10	7509	7394	21600	7443	(58.4–128.8)	**7382**	7442.47	13.68	**7382**	7421.33	13.96	**7382**	7442.63	11.84	**7382**	7394.13	53.11	**7382**	**7383.07**	32.12
ALM200-1	200	5	7467	7072	21600	7246	(251.7–527.9)	7079	7220.67	60.63	7092	7156.57	49.41	7123	7217.4	58.72	7061	7131.1	67.35	**7021**	**7081.8**	111.56
ALM200-2	200	5	7680	7225	21600	7367	(247.9–523.2)	7279	7383.13	64.09	7170	7266.7	49.79	7272	7348.47	58.61	7201	7231.77	84.20	**7124**	**7199.1**	93.74
ALM200-3	200	5	8217	7522	21600	7797	(210.2–518.7)	7547	7704.33	58.29	7525	7571.33	50.73	7545	7622.33	58.66	7495	7552.43	73.05	**7429**	**7513.33**	95.94
ALM200-1	200	10	10391	9615	21600	9595	(201.9–509.7)	9605	9798.33	46.08	9599	9725.2	39.16	9572	9806.47	49.82	9574	9769.07	60.18	**9550**	**9693.4**	84.52
ALM200-2	200	10	10238	9847	21600	9792	(182.2–509.7)	10056	10235.13	45.64	9939	10147.73	40.91	9824	10102.5	49.99	**9788**	9960.17	87.26	9822	**9953.87**	90.65
ALM200-3	200	10	10533	10018	21600	9955	(231.4–515.1)	9996	10224.2	48.17	9989	10102.73	42.12	9933	10122.5	50.51	9926	10116.73	92.18	**9880**	**9995.97**	77.81
ALM300-1	300	5	9871	8933	21600	9520	(438.5–1193.4)	9027	9216.67	149.5	8903	9021.57	115.45	9035	9152.43	135.43	8901	**8990.4**	196.62	**8869**	9008.67	198.25
ALM300-2	300	5	10532	9326	21600	9276	(443.2–1185.3)	8823	8989.1	154.55	8739	8845.9	109.56	8813	8910.37	136.45	8720	**8814.1**	179.99	**8688**	8817.93	205.42
ALM300-3	300	5	10887	9501	21600	10016	(506.0–1195.9)	9173	9327.8	143.5	9091	9158.97	121.82	9096	9206.07	136.56	9081	**9113.4**	204.98	**9021**	9115.47	232.12
ALM300-1	300	10	13899	12838	21600	12771	(474.7–1237.0)	12484	12773.37	116.58	12406	12612.57	93.3	12355	12545.37	115.62	12344	12545.33	140.42	**12186**	**12322.1**	204.74
ALM300-2	300	10	13210	12375	21600	12535	(445.5–1224.6)	12152	12371.27	124.16	12029	12223.37	97.22	12051	12177.9	117.86	12028	12271.2	211.49	**11774**	**11961.63**	172.7
ALM300-3	300	10	13792	13116	21600	12715	(480.8–1230.6)	12567	12844.03	120.99	12536	12644.93	93.18	12431	12610.03	119.09	12396	12577.07	329.37	**12252**	**12386.57**	203.69
ALM400-1	400	5	12487	12 487	21600	10811	(912.3–2503.2)	10439	10657.9	284.98	10321	10437.2	222.37	10421	10541.37	250.26	10311	**10349.8**	217.42	**10183**	10360.23	449.49
ALM400-2	400	5	13877	13877	21600	11090	(906.1–2374.0)	10511	10708.83	277.68	10350	10519.23	230.61	10468	10606.07	246.36	10343	**10403.87**	452.24	**10327**	10494.53	382.38
ALM400-3	400	5	12379	12379	21600	11098	(1022.7–2370.4)	10328	10534.3	280.26	10115	10266.97	222.5	10218	10346.57	249.08	10110	**10167.27**	244.18	**10073**	10264.13	382.5
ALM400-1	400	10	17309	17309	21600	15788	(811.4–2498.2)	14647	15008.8	198.4	14339	14580.1	174.77	14191	14527.33	212.39	14164	14548.13	348.21	**14032**	**14331.3**	185.42
ALM400-2	400	10	16595	16595	21600	15318	(850.3–2499.0)	14411	14765.27	219.14	14156	14411.17	167.49	14199	14322.4	212.9	14116	14280.37	180.30	**13765**	**14112**	161.27
ALM400-3	400	10	16439	16439	21600	15214	(819.8–2496.4)	14025	14474.77	217.88	13937	14140.6	163.32	13880	14111.03	214.22	13871	14161.07	288.54	**13552**	**13884.63**	200.15
ALM400-1	400	20	21339	21339	21600	20000	(832.0–2321.0)	20141	20672.67	188.08	19813	20217.93	154.79	19676	20096.13	211.63	19630	**19738.77**	536.92	**19559**	19817.73	312.13
ALM400-2	400	20	21299	21299	21600	20013	(719.9–2385.2)	19976	20386.53	163.43	19853	20080.87	148.05	19557	19879.4	208.13	19504	**19593.97**	517.66	**19442**	19665.47	245.36
ALM400-3	400	20	22049	22049	21600	20289	(787.1–2354.3)	19790	20584.6	162.98	19722	20137.93	149.69	19600	19980.57	204.68	19565	**19650.07**	443.22	**19396**	19691.77	207.68
ALM500-1	500	5	14626	14626	21600	11305	(1652.3–5338.9)	11316	11578.87	527.29	11119	11376.23	428.05	11332	11420.33	395.11	11105	**11143.5**	723.75	**11086**	11317.97	608.25
ALM500-2	500	5	14039	14039	21600	12405	(1541.0–5292.5)	11771	11978.53	518.53	11504	11770.1	479.46	11591	11748.33	390.94	11496	11577.9	959.97	**11461**	11663.23	719.06
ALM500-3	500	5	13251	13521	21600	11730	(594.7–5257.6)	11514	11708.23	574.44	11243	11472.77	465.8	11418	11537.7	390.07	**11224**	**11277.6**	871.69	11233	11414.13	646.34
ALM500-1	500	10	19342	19342	21600	17415	(1260.3–5433.9)	15688	15963.8	365.08	15258	15563.63	302.69	15490	15608.03	337.12	15258	15558.83	798.99	**15028**	**15310.97**	330.28
ALM500-2	500	10	18138	18138	21600	17653	(1382.3–5444.3)	16174	16492	368.16	15807	16205.27	309.73	15867	16044.43	333.33	15801	16065.97	402.96	**15586**	**15876.23**	290.88
ALM500-3	500	10	18269	18269	21600	16539	(1318.7–5376.9)	15931	16280.83	357.59	15628	16030.67	278.07	15608	15795.63	335.29	15599	15863.53	708.46	**15318**	**15556.33**	343.75
ALM500-1	500	20	24999	24999	21600	22346	(1302.1–5384.2)	21510	22252.47	278.13	21327	21844.67	227.9	21144	21494.03	328.41	21110	21243.13	550.99	**20781**	**21226.5**	265.94
ALM500-2	500	20	24823	24823	21600	22285	(1184.3–5285.6)	22018	22618.67	315.11	21758	22332.87	237.27	21603	22015.13	322.68	21540	**631.7**	491.95	**21340**	21740.5	260.01
ALM500-3	500	20	25468	25468	21600	22643	(1126.2–5320.0)	22281	22954.13	286.83	22100	22531.93	247.41	21865	22137.4	326.77	21792	**1918.83**	654.65	**21451**	21951.23	223.75
					20624.76					188.87			155.43			175.92			314.19			229.04

Table 5.4: Results of VNS^{I} against state-of-the-art approaches on E250 data-set

Instance	d	gen-0			gen-1			HSSGA			hABC			ILS			PSA			VNS^{I}		
		Best	*Avg*	*ATET*	*Best*	*Avg*	*ATET*	*Best*	*Avg*	*ATET*	*Best*	*Avg*	*ATET*	*Best*	*Avg*	*ATET*	*Best*	*Avg*	*ATET*	*Best*	*Avg*	*ATET*
E250.1	3	13.07	13.24	874.13	12.45	12.76	918.83	14	14.25	117.98	12.03	12.12	115.23	12.25	12.38	111.19	12.02	**12.03**	517.27	**12**	12.13	566.41
E250.2	3	12.85	13.01	516.04	12.6	12.86	871.47	13.42	14.03	99.35	**11.68**	11.8	111.62	11.99	12.12	110.47	**11.68**	**11.75**	467.21	11.74	11.84	493.61
E250.3	3	12.85	12.97	681.29	12.42	12.76	852.3	13.33	14.07	107.92	**11.79**	11.89	109.15	12.02	12.16	111.83	**11.79**	**11.84**	469.74	**11.79**	11.9	497.63
E250.4	3	12.78	13.04	605.41	12.86	13.02	850.86	14.06	14.25	112.64	12.11	12.19	108.82	12.38	12.5	110.89	12.1	**12.12**	534.78	**12.08**	12.22	509.61
E250.5	3	12.89	13.06	817.51	12.74	12.88	789.27	13.44	14.14	112.36	11.87	11.97	103.26	12.17	12.29	112.21	**11.86**	**11.9**	433.84	11.92	12.08	458.15
E250.1	5	16.45	16.67	650.4	16.65	17.25	1013.87	16.43	17.12	85.6	15.05	15.2	80.88	15.1	15.28	92.87	15.01	15.13	309.96	**14.76**	**14.94**	298.54
E250.2	5	15.55	15.83	574.67	16.04	16.35	1013.7	16.19	16.76	72.48	14.76	14.89	84.57	14.81	15.02	91.18	14.75	14.9	511.73	**14.56**	**14.7**	303.58
E250.3	5	17.08	17.37	633.26	15.36	15.73	992.89	16.14	16.87	81.36	14.6	14.78	80.24	14.65	14.95	91.9	14.58	14.78	345.43	**14.37**	**14.57**	407.74
E250.4	5	15.94	16.34	691.03	16.22	16.64	1031.76	17.01	17.28	75.45	15.12	15.36	79.85	15.33	15.5	91.96	15.12	15.33	388.72	**14.99**	**15.17**	359.14
E250.5	5	15.88	16.08	506.76	16.22	16.44	982.03	16.43	17.08	75.93	14.96	15.14	86.6	15.13	15.31	91.59	14.95	15.1	438.67	**14.87**	**15**	314
E250.1	10	23.35	23.82	683.07	23.81	24.22	1037.82	21.42	22.36	67.1	20.46	20.73	77.04	20.4	20.7	78.15	20.33	20.5	222.39	**20.07**	**20.49**	72.25
E250.2	10	22.3	22.67	613.44	22.65	24.09	982.23	21.46	22.4	62.91	20.39	20.73	65.39	20.5	20.84	78.28	20.32	**20.39**	307.72	**20.04**	**20.39**	88.23
E250.3	10	21.75	22.21	406.26	22.41	22.93	984.35	21.22	21.97	66.63	20.04	20.47	72.95	20.08	20.39	77.97	20.01	**20.11**	314.67	**19.81**	20.17	63.3
E250.4	10	22.74	24.3	710.56	23.17	24.01	1005.81	22.09	22.82	64.31	20.65	20.97	63.4	20.56	20.98	79.22	20.5	20.66	224.71	**20.3**	**20.65**	78.83
E250.5	10	22.49	23.48	737.53	23.27	23.68	985.97	21.6	22.54	62.3	20.32	20.57	70.63	20.39	20.76	78.89	20.28	**20.39**	268.16	**20.16**	**20.39**	77.28
				646.76			954.21			84.29			87.31			93.90			383.66			305.89

represent the following: for VNS^O, the best solution value (*Best*); for BC, the best solution value (*Best*) and the total execution time (TET); for min-GA*, both the best value (*Best*) and the average total execution time (TET) over 64 runs; and for HSSGA, hABC, ILS, PSA, and VNS^I, each column includes the best value across 30 runs (*Best*), the average value over 30 runs (*Avg*), and the average total execution time ($ATET$) over 30 runs. The last column represents the average of ATET over considered number of instances for each approach.

The column structure in Table 5.4, which is for Euclidean instances, remains largely consistent with Tables 5.2–5.3, with the exception that instead of VNS^O, BC, and min-GA*, it reports results for two min-GA variants (*gen-0* and *gen-1*). Each of these has three columns representing Best, Avg, and ATET based on 64 runs. In all tables, the best results are highlighted in bold for emphasis.

The results presented in Tables 5.2–5.3 demonstrate that the proposed VNS^I consistently outperforms the existing VNS^O across CRD, SYM, and ALM data-sets, particularly in terms of solution quality. When compared to BC approach, VNS^I achieves better results on 5 instances, performs equally on 30, and is less effective on 19 within CRD and SYM data-sets based on the *Best* values; however, VNS^I is superior to BC on ALM instances (instances of larger size) except two instances. When compared to min-GA*, VNS^I achieves better results on 8 instances, performs equally on 28, and is less effective on 18 within CRD and SYM data-sets based on the *Best* values; however, VNS^I is superior to min-GA* on all ALM instances except on one instance in which min-GA* finds better result and on two instance in which min-GA* performs as same as VNS^I. Also, VNS^I outperforms *gen-0* and *gen-1* variants of min-GA on all instances of E250 data-set.

Table 5.5 presents a comprehensive comparison between the proposed VNS^I and the four most recent metaheuristic techniques: HSSGA, hABC, ILS, and PSA. The comparison spans three benchmark data-sets – CRD+SYM, E250, and ALM – based on the results compiled from Tables 5.2 through 5.4. For each approach, we considered two critical performance indicators: the *Best* solution obtained over 30 runs and the *Average* solution quality over 30 runs. In Table 5.4, the comparative symbols are defined as follows:

- "+" signifies that VNS^I achieved a superior result.

□ "=" indicates that both methods showed comparable results,

□ "–" denotes that VNS^I yielded a comparatively inferior result.

Table 5.5: Comparison of VNS^I with existing PSA, ILS, hABC, and HSSGA approaches across all four datasets in terms of *Best* and *Avg* values.

Approaches	**CRD+SYM**						**E250**						**ALM**					
	Best			Avg			Best			Avg			Best			Avg		
	+	=	–	+	=	–	+	=	–	+	=	–	+	=	–	+	=	–
VNS^I vs PSA	4	32	18	14	3	37	12	1	2	7	2	6	29	4	3	21	0	15
VNS^I vs ILS	14	27	13	26	0	28	15	0	0	15	0	0	33	3	0	36	0	0
VNS^I vs HABC	10	27	17	18	0	36	12	1	2	10	0	5	33	3	0	35	0	1
VNS^I vs HSSGA	8	30	16	24	1	29	15	0	0	15	0	0	33	3	0	36	0	0

The results indicate that VNS^I performs particularly well on ALM and E250 data-sets. Compared to HSSGA, VNS^I yields better results across all instances on ALM and E250 data-sets. However, on the CRD and SYM data-sets, HSSGA performs better in terms of average values, though both methods show comparable performance when considering only the best results.

Against hABC, VNS^I does not perform as well on CRD+SYM, particularly in terms of average values, it demonstrates strong performance on the ALM and E250 data-sets across nearly all instances, both in best and average values.

When compared to ILS, VNS^I clearly gives better results on both E250 and ALM, with consistent improvement in all cases. On CRD+SYM, the results are more mixed, but VNS^I still performs better or equally in many instances.

Compared to PSA, PSA performs better on CRD+SYM data-sets. On E250 data-set, VNS^I performs better in terms of best values and gives comparable average results. Its performance on ALM is especially better, outperforming PSA in both best and average values in most cases.

In terms of computational time, authors of [44] allocated significantly more runtime to PSA approach compared to HSSGA, hABC, and ILS, enabling PSA to achieve improved results, especially on the ALM and E250 datasets. In contrast, VNS^I requires less computational time than PSA while consistently delivering superior solution quality on the larger instances, notably ALM and E250. This highlights the efficiency and effectiveness of our approach in handling large problem instances.

Overall, VNS^I dominates over all existing approaches on ALM and E250 data-sets. Notably, VNS^I finds new values on 43 instances out of 105 instances within all four data-sets, highlighting its ability to search new values in the search space.

5.3.5 Statistical test

To assess the relative performance of the proposed VNS^I, statistical significance tests are conducted using the best value (*Best*) and average value (*Avg*) values obtained across all data-sets. For this Wilcoxon's signed rank test [69] with a significance criterion of 0.05 is used. The statistical results are grouped by data-set type: CRD+SYM, ALM, and E250 and reported in Table 5.6-5.8.

Based on statistical results reported in Table 5.6-5.8, it is found that VNS^I shows statistically significant difference with PSA in terms of *Best* on all data-sets. In terms of *Avg*, although statistical tests suggest no significant difference in average values (*Avg*) between VNS^I and PSA on ALM and E250 data-sets, VNS^I performs superior to PSA on ALM and E250 data-sets and finds new values on many instances.

On CRD and SYN data-sets (instances of smaller size), statistical tests suggest no significant difference in *Best* and *Avg* between VNS^I and HSSGA, hABC and ILS except on hABC in terms of *Avg*. However, VNS^I shows statistically significant difference with HSSGA, hABC and ILS in terms of *Best* and *Avg* on ALM and E250 data-sets.

Table 5.6: Statistical results for CRD+SYM data-sets

Approaches	*Best*		*Avg*	
	Significant	p-value	Significant	p-value
VNS^I vs. PSA	Yes	0.00988	Yes	0.00042
VNS^I vs. ILS	No	0.86502	No	0.09102
VNS^I vs. hABC	No	0.16452	Yes	0.00018
VNS^I vs. HSSGA	No	0.23404	No	0.38978

Table 5.7: Statistical results for ALM data-set

Approaches	*Best*		*Avg*	
	Significant	p-value	Significant	p-value
VNS^I vs. PSA	Yes	0.00001	No	0.13888
VNS^I vs. ILS	Yes	0.00001	Yes	0.00001
VNS^I vs. hABC	Yes	0.00001	Yes	0.00001
VNS^I vs. HSSGA	Yes	0.00001	Yes	0.00001

Table 5.8: Statistical results for E250 data-set

Approaches	*Best*		*Avg*	
	Significant	p-value	Significant	p-value
VNS^I vs. PSA	Yes	0.00424	No	0.50926
VNS^I vs. ILS	Yes	0.00064	Yes	0.00064
VNS^I vs. hABC	Yes	0.00424	Yes	0.01046
VNS^I vs. HSSGA	Yes	0.00064	Yes	0.00064
VNS^I vs. *gen-0*	Yes	0.00064	Yes	0.00064
VNS^I vs. *gen-1*	Yes	0.00064	Yes	0.00064

5.3.6 Parameter tuning overhead: VNS^I vs. PSA

PSA [44] is a recent and competitive metaheuristic that has shown better performance compared to earlier methods, particularly on instances of larger size (ALM and E250 data-sets). However, when evaluating the practicality of our proposed VNS^I in relation to PSA, especially in terms of parameter tuning overhead, a clear advantage emerges. PSA requires tuning 12 parameters, whereas VNS^I relies on only 2. This significant reduction in parameter count not only simplifies the tuning process but also reduces computational effort, enhances reproducibility, and facilitates easier deployment in real-world applications or on unseen instances. Moreover, the strong performance of VNS^I – particularly on larger benchmark datasets such as ALM and E250 – demonstrates its robustness and practical effectiveness, even when compared to more heavily parameterized PSA approach.

Chapter 6

Conclusions and Future Research Directions

Over recent decades, the field of $\mathcal{NP}$-hard combinatorial optimization problems (COPs) has experienced remarkable expansion. These problems are not only theoretical challenging but also possess diverse practical applications, manifesting in diverse real-world applications within domains like operations research and computer science.

It has become increasingly evident that relying solely on basic metaheuristic techniques often results in poor solution quality for $\mathcal{NP}$-hard COPs. As a result, researchers are continuously striving to discover high-quality solutions in a resonable computational time. This effort often involves designing new strategies or ingeniously adapting existing ones, specifically tailored to the unique characteristics of a given $\mathcal{NP}$-hard COP – to develop robust and effective metaheuristic techniques. Such new strategies often come from a deep understanding of the problem's inherent structure and domain-specific knowledge. Such insights can be strategically integrated into key components of a chosen metaheuristic framework, including the way solutions are represented (encoded), how initial solutions are generated, and the design of specialized operators (such as neighborhood operators). Moreover, these problem-specific insights are instrumental in developing effective local search procedures.

This thesis focused on four distinct $\mathcal{NP}$-hard combinatorial optimization problems - [1] minimum edge-dilation K-center (MEDKC) problem, [2] tree t-spanner problem

(Tr_t-SP), [3] single machine scheduling problem with periodic machine availability (PMASMSP), and [4] min-degree constrained minimum spanning tree (MDMST) problem – through three metaheuristic techniques such as artificial bee colony algorithm, iterated local search, and variable neighborhood search. These problems find applications in routing schemes within computer networks; in various production environments where maintenance activities play a critical role in ensuring efficiency, safety, and improved productivity; distributed systems, communication networks, network design, robotics, and computational biology, particularly in building network synchronizers and reconstructing phylogenetic trees; and VLSI circuit design, data clustering, and telecommunications.

Our research work focused on designing robust and effective metaheuristic techniques tailored to four distinct $\mathcal{NP}$-hard combinatorial optimization problems. The research emphasizes the use of problem-specific knowledge in key components of metaheuristic techniques – such as encoding, solution initialization, and neighborhood operator – and appropriate operators to enhance performance. Where appropriate, suitable local search strategies or their extensions are integrated within the metaheuristic frameworks. The overarching goal is to ensure that these components operate cohesively to effectively explore the search space and find high-quality solutions within reasonable computational time.

In this work, we proposed several metaheuristic techniques designed for specific $\mathcal{NP}$-hard problems: artificial bee colony (ABC) algorithm and iterated local search (ILS) for the MEDKC problem; iterated local search (ILS) and variable neighborhood search (VNS) for the Tr_t-SP; an iterated local search (ILS) approach for the PMASMSP; and an improved variable neighborhood search (VNS^{I}) approach for the MDMST problem.

The following paragraphs outline the key contributions presented in each chapter and highlight potential directions for future research.

In chapter 2, we present two metaheuristic techniques tailored to the MEDKC problem. The first one is an artificial bee colony (ABC) algorithm that applies two neighborhood operators in a mutual exclusive way specifically tailored to this problem. The first neighborhood operator leverages the insight that if a center vertex frequently appears in high-quality solutions, it is likely to be a promising component in other potential solutions of employed bee population as well, whereas the second neighborhood is based on the swapping of center vert(ex/ices) with non-center vert(ex/ices) in a mixed strategies of greedy and random approach. The synergistic coordination of all components within ABC algorithm contributes to achieving solutions of high quality. The second one is an iterated local search (ILS) in which a local search is proposed that effectively exploits the local region around the perturbed solution, where perturbed solution is generated through swapping a random center vertex with a random non-center vertex of the current solution.

Similar to two existing genetic algorithms (GA1 and GA2) [34], we have also used the same two set of instances which are available in public domain. These benchmark sets comprises 49 BX and 42 BB graphs. The BX set has 7 4-regular graphs which have different sizes of vertices from the set $\{25, 50, 75, 100, 200, 400, 1000\}$. The second BB set has 7 graphs: 5 edge-weighted graphs of grid type whose sizes are $< 15 \times 15, 33 \times 33, 45 \times 5, 50 \times 50$ and $100 \times 10 >$ and two large Steiner tree benchmark instances, i.e., steinc5 and steind5 that have 500 vertices and 1000 vertices. Each instance is examined for different values of K whose values are $\in \{2, 3, 4, 5, 10, 15, 20\}$. Our proposed approaches outperformed two existing genetic algorithms (GA1 and GA2) and a polynomial-time approximation algorithm (Konemann_AL) [35] across various performance metrics. In terms of best solution values, ABC algorithm showed improvements over GA1 on 18 instances and GA2 on 61 instances, while ILS showed improvements on 20 and 60 instances respectively. When average solution quality was considered, ABC algorithm surpassed GA1 on 68 instances and GA2 on 86, whereas ILS performed better on 65 and 69 instances respectively. Both our proposed approaches also consistently outperformed Konemann_AL on all 91 tested instances.

Furthermore, ABC algorithm and ILS discovered 17 and 18 new best-known solutions respectively. Observing on computational time (the average total execution time (ATET)), ABC algorithm and ILS are overall faster than GA1 and GA2.

Between ILS and ABC algorithm, ILS obtains superior best-known results on 11 instances, matches ABC algorithm on 78 instances, and performs worse on 2 instances. Regarding average solution quality, ILS shows improved performance on 26 instances, achieves similar results on 27, and yields lower quality solutions on 38 instances in comparison to ABC algorithm.

As a future work, the neighborhood operators used in ABC algorithm and operators such as perturbation operator and local search in ILS can be applied in other metaheuristic techniques specifically tailored for the MEDKC problem.

Chapter 3 presents two metaheuristic techniques tailored to the Tr_t-SP. The first proposed ILS relies on the problem-specific perturbation strategy involving two successive edge deletions and additions, a problem-specific local search improvement strategy, and an acceptance criterion that combines two strategies – where the second helps ILS escape deep local optima. The second proposed VNS method systematically explores three neighborhood structures based on problem-specific and employs a dedicated local search improvement strategy to refine solutions. Additionally, a restart criterion mechanism is incorporated to increase the chances of discovering new high-quality solutions.

The experimental evaluation was conducted on available 48 benchmark Euclidean graph instances (with 50 to 200 vertices and varying edge densities). The performance of the proposed ILS and VNS is evaluated against the existing ABC algorithm [36], as other metaheuristic techniques from the literature were not considered due to their results being consistently outperformed by the ABC algorithm.

When comparing VNS and ILS on 48 instances, VNS achieves better best-known values on 13 instances, matches performance on 10, and performs worse on 25 instances. With respect to average solution quality, VNS delivers better results on 23 instances, shows equal performance on 3, and falls behind on 22 instances when compared to ILS.

As a future work, both ILS and VNS offers simplicity with fewer parameters compared to the existing ABC algorithm, facilitating its application to new problem instances. Secondly, the core concepts, notably the utilization of two strategies in the acceptance criterion in ILS and a restart criterion in VNS, can be extended to related problems like the quadratic minimum spanning tree and minimum routing cost spanning tree problems. Thirdly, there are instances where the average solution quality over 10 runs differs from the best value obtained by ILS, implying that further improvement is still possible. For instance, the proposed problem-specific local search improvement strategy can be investigated with other metaheuristic techniques such as simulated annealing, memetic algorithm for the Tr_t-SP. Also, other metaheuristic techniques such as tabu search can be adapted for the Tr_t-SP.

Chapter 4 presents an iterated local search (ILS) approach tailored to the PMASMSP. In the proposed ILS, the perturbation procedure is based on swapping of jobs in different periods, a series of local search strategies involving 1-0 insertion, 1-1 swap, 1-2 swap, and 2-1 swap, all focused on minimizing the processing time in the last period, and acceptance criterion follows two step procedures in which the second step, particularly, is used to prevent ILS to be trapped into deep local optima.

The performance of the proposed ILS was evaluated against existing 19 approaches (15 constructive heuristic, 3 heuristics (NEW, NEW_FF and NEW_BF), and a MILP model) [37], and Genetic algorithm (GA) [38] on available benchmark instances. These instances are categorized into two class of benchmark instances called *LOW* and *MOD* instances generated by Paz and Jose [37]. Each class of instances has 14 categories, with the number of jobs falling within the range of {10, 20, 30, 40, 50, 60, 70, 80, 90, 100, 150, 200, 250, 300}. In each category, there are also 50 instances. Hence, each class of benchmark instances comprises of 700 instances.

On MOD instances, the performance of ILS is superior to all 19 approaches except three categories of instances with n = 10, 20, 30, where the performance of ILS is equivalent to that of MILP. When comparing with genetic algorithm, ILS performs better. In fact, ILS discovers new best results on 6 out of 14 categories of *MOD* instances. On LOW instances, the performance of ILS and GA are similar. Its computational time is particularly notable when compared to the top three among these 19 approaches [37] and GA [38], making it highly applicable in a production environment.

As a future work, ILS framework presented for the PMASMSP problem is versatile and can be applied to new instances of the problem. It is to be noted that ILS does not outperform existing GA in 3 out of 14 instance classes. This indicates that ILS can still get trapped in local optima, suggesting potential future work to explore other metaheuristic techniques combined with the local search strategies used in ILS for a better approach.

Chapter 5 presents an improved variable neighborhood search (VNS^I) approach tailored to the MDMST problem. In contrast to existing four VNS variants [39], VNS^I incorporates degree-aware edge selection within a multi-neighborhood framework comprising four structurally distinct neighborhoods (N_1–N_4). These are designed to prioritize edge removals that significantly influence the degree of vertex, thereby improving its ability to navigate the search space. Coupled with tailored two different local search strategies, this neighborhood design enhances both intensification and diversification capabilities. A restart criterion is incorporated to increase the chances of discovering new high-quality solutions.

The performance of the proposed VNS^I was evaluated against the best-performing result among the existing four VNS variants under the label VNS^O [39], branch and cut approach [40], the best-performing result among the three variants of genetic algorithm (min-GA*) – *gen-0*, *gen-1*, and *gen-2* [41], HSSGA [42], ILS, and hABC [43], and a population-based simulated annealing (PSA) [44] on available four standard benchmark datasets, namely CRD, SYM, ALM, and E250, covering 105 instances with varying graph sizes and structures. The proposed VNS^I consistently outperforms the existing VNS^O across CRD, SYM, and ALM data-sets, particularly in terms of solution quality. When compared to BC approach, VNS^I achieves better results on 5 instances, performs equally on 30, and is less effective on 19 within CRD and SYM data-sets based on the $Best$ values; however, VNS^I is superior to BC on ALM instances (instances of larger size) except two instances. When compared to min-GA*, VNS^I achieves better results on 8 instances, performs equally on 28, and is less effective on 18 within CRD and SYM data-sets based on the *Best* values; however, VNS^I is superior to min-GA* on all ALM instances except on one instance in which min-GA* finds better result and on two instance in which min-GA* performs as same as VNS^I. Also, VNS^I outperforms *gen-0* and *gen-1* variants of min-GA on all instances of E250 data-set. The results show that VNS^I performs best on the ALM and E250 data-sets, consistently outperforming HSSGA, hABC, and ILS on all instances. While it lags behind HSSGA and hABC in average values on CRD+SYM, it remains competitive in best results. Against ILS, VNS^I is superior on all E250 and ALM instances and performs comparably or better on CRD+SYM. Compared to PSA, on E250 data-set, VNS^I performs better in terms of best values and gives comparable average results. Its performance on ALM is especially better, outperforming PSA in both best and average values in most cases, while PSA performs better on CRD+SYM data-sets. Overall, VNS^I dominates on ALM and E250 data-sets, and discovers 43 new best values across 105 instances – demonstrating its capability to search new values in the search space.

In terms of computational time, authors of [44] allocated significantly more runtime to PSA approach compared to HSSGA, hABC, and ILS, enabling PSA to achieve improved results, especially on the ALM and E250 datasets. In contrast, VNS^I requires less computational time than PSA while consistently delivering superior solution quality on the larger instances, notably ALM and E250. Also, compared to the recent PSA approach [44] – which, despite its strong performance compared to earlier approaches on ALM and E250 data-sets at the cost of high computational time, requires tuning 12 parameters – VNS^I achieves competitive or superior results with only 2 parameters. This highlights its robustness, practical applicability, and effectiveness, especially for large-scale problem instances.

As a potential avenue for future work, the core design strategy of VNS^I can be adapted to tackle other $\mathcal{NP}$-hard combinatorial optimization problems involving degree constraints.

The research work presented in this thesis emphasizes that designing robust and effective metaheuristic techniques for $\mathcal{NP}$-hard COPs requires the importance of a deep understanding of both the problem domain and the behavior of these search strategies. Key factors influencing the success of a metaheuristic technique include the choice of solution encoding schemes, the selection and configuration of appropriate neighborhood operators, and the synergy between different components of a metaheuristic technique. Another important aspect is the integration of problem-specific knowledge into components such as initial solution generation, neighborhood operator design, and local search procedures. Achieving a balanced trade-off between exploration (diversification) and exploitation (intensification) is essential to avoid premature convergence and to guide the search toward high-quality solutions. These aspects are critical for designing a metaheuristic technique that is both robust and effective for NP-hard COPs.

Despite the contributions made in this thesis, there remain several unexplored areas whose further exploration of these could significantly advance the understanding of the intricate characteristics of a particular $\mathcal{NP}$-hard combinatorial optimization problem, thereby guiding the development of even better robust and effective metaheuristic techniques. These unexplored avenues offer substantial opportunities to further improve metaheuristic techniques for the solution of such challenging problems.

www.ingramcontent.com/pod-product-compliance
Lightning Source LLC
LaVergne TN
LVHW021736071126
851857LV00040B/1458
* 9 7 8 1 9 6 2 1 1 6 6 1 9 *